Hasret Dikici Bilgin

Working Street Children in Turkey and Romania

A Comparative Historical Analysis in the Context of New Poverty

VDM Verlag Dr. Müller

Impressum/Imprint (nur für Deutschland/ only for Germany)
Bibliografische Information der Deutschen Nationalbibliothek: Die Deutsche Nationalbibliothek
verzeichnet diese Publikation in der Deutschen Nationalbibliografie; detaillierte bibliografische
Daten sind im Internet über http://dnb.d-nb.de abrufbar.

Coverbild: www.purestockx.com

Verlag: VDM Verlag Dr. Müller Aktiengesellschaft & Co. KG
Dudweiler Landstr. 99, 66123 Saarbrücken, Deutschland
Telefon +49 681 9100-698, Telefax +49 681 9100-988, Email: info@vdm-verlag.de
Zugl.: Ankara, Middle East Technical University, Diss., 2006

Herstellung in Deutschland:
Schaltungsdienst Lange o.H.G., Berlin
Books on Demand GmbH, Norderstedt
Reha GmbH, Saarbrücken
Amazon Distribution GmbH, Leipzig
ISBN: 978-3-639-13036-2

Imprint (only for USA, GB)
Bibliographic information published by the Deutsche Nationalbibliothek: The Deutsche
Nationalbibliothek lists this publication in the Deutsche Nationalbibliografie; detailed
bibliographic data are available in the Internet at http://dnb.d-nb.de.

Cover image: www.purestockx.com

Publisher:
VDM Verlag Dr. Müller Aktiengesellschaft & Co. KG
Dudweiler Landstr. 99, 66123 Saarbrücken, Germany
Phone +49 681 9100-698, Fax +49 681 9100-988, Email: info@vdm-publishing.com
Ankara, Middle East Technical University, Diss., 2006

Printed in the U.S.A.
Printed in the U.K. by (see last page)
ISBN: 978-3-639-13036-2

TABLE OF CONTENTS

LIST OF TABLES

4

LIST OF FIGURES

ACKNOWLEDGMENTS

I would like to express my indebtedness to my supervisor Assoc. Prof. Dr. Ayşe Gündüz Hoşgör for her guidance, advice, criticism and encouragement. This study would not be completed without her invaluable support. Her guidance helped me to ask the right questions and think analytically.

I would also like to thank Prof. Dr. Yıldız Ecevit and Assoc. Prof. Dr. Erkan Erdil for their constructive criticisms which helped me to reevaluate my approach.

I should finally express my gratitude to my family: my father Rahmi Dikici, my mother Münevver Dikici, my sister Cennet Dikici, and my husband Halil Bilgin, for their love and support in every stage of this study.

6

CHAPTER I

INTRODUCTION

We are living in an age of increased opportunities, commodities and services. Generally speaking, we have more money to spend and more commodities are available to be bought. *1990 Human Development Report* acknowledges this trend and states that per capita income increased rapidly since 1980s, in rates higher than it was between 1965 and 1980 (UNDP 1999, 21). Life expectancy rates in the developing countries have increased by one third since 1960; world average rising to as high as 67 years (UNDP 2005, 222).

Given these developments, it might be claimed that it is a general progress in the historical course of human well-being; however, this progress has not been even, and many social groups could not benefit from the developments as much as the rest of people. Children are among these groups; and despite this general progress, child malnutrition and infant mortality rates have risen in many developing countries (UNDP 1990, 18). In the last few decades, it is even the case that child poverty increased at a higher rate than overall poverty (Bradshaw 2000, 223-250). In addition to the increased child poverty, the children are exposed to social exclusion (UNICEF 2005).

These findings on children indicate a dilemma. On the one hand, children might be claimed to benefit from increased life standards. For instance, number of child labourers in the world is estimated to decrease by 11% from the year 2000 to 2004 (IPEC 2006). A further 24% decrease occurred in the number of children engaged in hazardous labour (IPEC 2006). On the other hand, there are still millions of children working in heavy occupations – 218 million child labourers by 2004 figures (IPEC 2006); more than 100 million is estimated to be sexually exploited, subject to slavery or street life around the world (Kittredge and Nielsen 2003, 19). It is also estimated that there are tens of millions of street children (UNICEF 2005, 40). Adding those children only working on the street, not living on the street, it might be claimed that this number is far higher.

7

Among these children, working street children constitute the topic of this study. At first sight, one might think that working street children are relatively in better conditions than, for instance, children in bondage or working in mines, judging by the scenes of children selling napkins and chewing gums on the street in daytime. On the contrary, in my opinion, those children we see in the city centers in daytime are just a part of the whole, and working on the street includes many more jobs; some of them being quite dangerous for children. Many working street children today have become "invisible" (in the sense of violation and ignorance of their right and exclusion) from the gaze of people and even from the gaze of the governments (UNICEF 2005, 35). *State of the World's Children: Excluded and Invisible* report refers to street children in this group, no matter whether they are living or working on the street (UNICEF 2005, 36). Moreover, working street children are distinguished from many other children in exclusion, in that, although they are "physically visible" – unlike, for instance, trafficked children – they are "among the most invisible and, therefore, hardest children to reach with vital services, such as education and health care, and the most difficult to protect" (UNICEF 2005, 40).

Acknowledging the rapid progress in human well-being, particularly since 1980s; emergence of working street children is a quite interesting and seemingly paradoxical situation. Existence of these children in different countries, not only among the developing countries, but also in the highly industrialized countries (UNICEF 2005, 41), in my opinion, further complicates the discussion of progress and development.

This study aims to gain insight into the causes for emergence and expansion of working street children. Acknowledging the fact that working street children is not a notion specific to Turkey and almost all cities of the developing countries have working street children (UNICEF 2005), a comparative analysis with another country is preferred. In this context, Romania, as a country with working street children is chosen. Why Romania is chosen in order to be compared with Turkey in terms of working street children have a number of reasons. First of all, as mentioned above, there is existence of working street children in this country. Secondly, picking up a country, whose political, economic, social and cultural characteristics are relatively different from Turkey, but still having working street children, might broaden understanding on the causes behind emergence of working street children and might help questioning different perspectives which, otherwise, I might not notice by looking only at the Turkish case. Romania, as a country from the former Eastern Bloc, might have provided that difference. Thirdly, Turkey and Romania are part of the International Programme for the Elimination of Child Labour (IPEC) under International Labour Organization (ILO). By IPEC, working street children are identified among the worst forms of child labour in both Turkey and Romania (IPEC 2003, 70) and research is carried out on working street children as one of the worst forms of child labour in different countries. Turkey participated in IPEC in 1992 (Akşit et al. 2001, ix) and IPEC is practically in Romania since 2000 (Alexandrescu 2002, ix). Therefore, the system of the combating with the problem of working street children might be considered similar; and an important number of research in the two countries are carried out in this framework increasing the opportunity for comparison.

In terms of the causes of emergence of working street children, poverty is scrutinized. Poverty is identified as the main cause of child labour in general (Basu and Tzannatos 2003, 16). However, there are references to 1980s and 1990s in terms of the emergence of working street children. For instance, in *Turkey, Working Street Children in Three Metropolitan Cities: A Rapid Assessment* (Akşit et al. 2001, 64), it is stated that "...the majority of families of children working in the streets came to the urban centers after 1985..." Kahveci et al., in the study on shoe-shine boys in İzmir mentions that only 10% of the families of these children are in the city for more than 15 years (Kahveci et al. 1996, 41), which implies that the families migrated to this city after 1980s. In the research on Romania, I did not come across such a statement; nevertheless, the studies refer to the decrease of living standards since 1990 when they are identifying the factors of working street children and child labour (Alexandrescu 2002, 3; Save the Children Romania and IPEC 2003, 8). Thus, we can trace emergence of working street children, at least in Turkey and Romania, to the late 1980s and 1990s.

In this context, neither poverty nor child labour is new. There are studies on child labour, particularly in the 19th century in the industrial settings (Basu and Tzannatos 2003, 3). Almost half of the textile workers in the United Kingdom are claimed to be under 18 years old in 1839 (Atauz 1989a, 1). Depending on the references of mentioned above; however, it might be claimed that working street children as a form of child labour is a considerably new concept. Therefore, what has changed in the last two decades so that working street children emerged? The causes behind emergence of working street children as well as the scope and dimensions constitute the first research question of this study.

As mentioned above, in both Romania and Turkey, the studies on working street children refer to similar dates (Akşit et al. 2001; Alexandrescu 2002). What has changed with poverty so that working street children emerged as a new form of poverty and child poverty and what might be the reasons behind emergence of working street children in these two countries at similar time periods? This question constitutes the second research question.

In terms of methodology, the main social research methodology employed during the study is Comparative Historical Analysis. Comparative Historical Methodology is suitable for questions addressing major social changes, particularly why some social events have taken place in some societies, but not in others (Neuman 2000, 404). The method is also useful in analyzing the causes behind the social changes (Mahoney 2004, 82). In comparative historical studies, historical and cultural characteristics are combined with theoretical framework and case studies are compared in historical course. In short, an event is examined for a number of countries within a historical framework (Neumann 2000, 406-407). In this respect, what has changed in terms of poverty so that working street children emerged as a form of working street children and as a reflection of child poverty; and why emergence of working street children coincide in Turkey and Romania in terms of time periods might be utilized with this method.

The study begins with a theoretical framework on poverty and working street children. Initially, situation of children, in general, is provided. Specific attention is devoted to the increase in child poverty at a higher rate than the poverty in general. Then poverty is discussed. And it is argued that poverty has gone through transformation in late 1980s and early 1990s, in accordance with the increased influence of global dynamics and neo-liberal paradigm. Within the framework of neo-liberal paradigm and globalization, many countries have taken measures of market reform in order to decrease public expenditures to pay their debts to international monetary organizations and increase the competitiveness of their economies (Cornia 2001). In line with the liberalization and privatization reforms and decreased governmental role; the economic structure changed, many social policies were withdrawn and living standards fell (Ecevit 1998). In this chapter, it is argued that a new form of poverty emerged under these circumstances called *New Poverty,* in which relative poverty, inequality and social exclusion are significant dimensions (Esping-Andersen 1999, Buğra and Keyder 2003). Child poverty is claimed to be affected from the transformation of poverty, and working street children emerged as a new form of child labour in this context. The chapter is concluded with terminology on working street children, since there are various concepts on the children working and/or living on the streets and the concepts should be clarified in order to make a discussion on them.

Chapter III begins with the review of existing studies on child labour and working street children in Turkey. Then, child poverty in Turkey after 1980s is discussed in the context of transformation of poverty and child poverty. Next, emergence of working street children as a new form of child poverty and child labour is discussed together with its causes. The scope and dimensions of working street children is considered finally together with the response measures taken against the problem.

In Chapter IV, working street children in Romania is discussed following a similar outline with the chapter on Turkey. After the brief literature review, child poverty, factors for child poverty – international and national factors, emergence of working street children, the causes behind working street children issue in Romania and the response measures are provided.

Chapter V compares working street children in Turkey and Romania trying to locate the comparative analysis within the context of *New Poverty*. Situation of children in Turkey and Romania is compared initially, followed by the comparison on working street children. The causes, scope and dimensions of working street children is compared for two countries. This chapter is concluded with an analysis part. The analysis compares the children under highest risks on the children and other dimensions of the issue. In the concluding chapter of this study, the findings of the comparison in the previous chapter are discussed within the context of *New Poverty* in an attempt to integrate the findings to the theoretical framework.

CHAPTER II

WORKING STREET CHILDREN IN THE CONTEXT OF
CHILD POVERTY AND NEW POVERTY

2.1 The Contemporary Situation of the World's Children

Children are the future assets of a society. How a nation takes care of the well-being of its children is a significant indicator of how that society concentrates on the future well-being of its citizens in general. Child poverty has consequences on children's mental and physical development, health and survival, educational success and job prospects, incomes and life expectancy (Vleminckx and Smeeding 2001, 75). Accordingly, a child growing in such circumstances have lower chances of having a better life in adulthood and contributing to the prosperity of the society. Therefore, discussing the effects of poverty on people, child poverty deserves a privileged attention.

Despite high economic growth and the improvements in the living standards in the industrialized nations throughout the second half of 20th century, an important portion of children are still living in poor families with health and growth difficulties. The previous social policies, at least to an extent, have managed to reduce poverty among older people, however, their success in reducing poverty for younger people remained much more limited. In the last two decades, child poverty rose substantially in the developing world, it is even witnessed re-emergence of high child poverty among the populations of Western industrialized nations (Vleminckx and Smeeding 2001, 1). Poverty has increased in the previous decades as mentioned before; however, studies show that child poverty has increased more (Bradshaw 2000, 223-250).

UNICEF's report on the situation of children all over the world, namely *The State of the World's Children 2005: Childhood under Threat* provides striking facts about child poverty and well-being. According to a UNICEF report, around 1 billion of children over a total of 2.2 billion children in the world live in poverty, which means every second child is exposed to poverty (UNICEF 2004). One of three children lack adequate shelter, one of five children has no access

13

to safe water and one in seven children lacks health services. An important portion of children is out of school, most of them being girls, many of them do not have access to communication and information, child mortality rates are strikingly high to the extent that children dieing before 5 is equal to the number that survive even in some rich countries (UNICEF 2004, 2). A key finding of the report is that many of the children in extreme poverty live in the countries with relatively high incomes (UNICEF 2004, 23). Gender discrimination is also identified as a significant factor in child poverty (UNICEF 2004, 23-24). The fact that the statistics do not exclude the children in the comparably developed countries gives an idea for the scope of child poverty.

Another UNICEF report reveals that the relative child poverty, which is defined as exclusion of children from activities and advantages their peers have access although their physical needs might be at least minimally sustained, (UNICEF Innocenti Research Center 2000, 2), is highest in the United Kingdom, Italy, United States and Mexico; while absolute child poverty, referring to deprivation from basic needs due to income poverty of their families, is highest in Spain, Czech Republic, Hungary, and Poland. The report finds positive relation between child poverty and parental unemployment, single-parent situation, working poor families, and low governmental social expenditure (UNICEF Innocenti Research Center 2000, 2).

High levels of child poverty re-emerged even in the rich nations and became widespread and visible so that European Union member states had to put an additional article on the child poverty in their Lisbon Council on the new European Social Agenda, with the aim of taking steps against poverty in general, however paying special concern for child poverty together with other groups under risk (Vleminckx and Smeeding 2001, 3).

According to the Luxembourg Income Study covering 25 industrialized countries all over the world (http://lis.ceps.lu), there is significant variation in terms of child poverty between countries at similar levels of development. The result of the statistics implies that the factors other than income in the form of market transfers, specifically speaking, family structure (demographic transformation) and state transfers play role in terms of child poverty (Bradbury et al. 2001, 11).

These findings indicate a dilemma in the historical course of human well-being. On the one hand, economy expanded and technology has advanced so that many goods and services in nutrition, health and education, which were previous luxuries for ordinary people, have become items of daily access for the mass population. Yet, poverty could not be eliminated totally, the level some people suffer poverty even rose, and the recent statistics show that child poverty has raised more than poverty increased in general (Bradshaw 2000, 223-250).

Co-existence of generally increased life standards and increased child poverty both in developed and developing countries is a question in itself. As mentioned in the first chapter, poverty is not a new problem humanity is faced with, neither is child poverty. The children of poor families have used to suffer from similar problems with their families, and most of them had to work either on the household farms or outside in order to contribute to the family budget. However, first of all, it was relatively easier for them to find jobs that might go hand in hand with the school, usually jobs in the industrial settings, small workshops and farms. This does not indicate far better living conditions, but at least upward mobility through education or training at work environment used to be more available. Nevertheless, today many children have to drop out the school in order to work, and in the majority of these cases, they are the sole breadwinners in their families (Kahveci et al. 1996, 38). From this perspective, working street children provide a distinguished example for the contemporary form of child poverty.

The main aim of this study is to gain insight into the working street children in such a theoretical framework. In this chapter, in order to understand why working street children emerged as a contemporary form of child poverty, the dynamics behind the process that led to the emergence of working street children are scrutinized. Briefly speaking, poverty in general is discussed. It is argued that poverty itself has gone under a transformation process that resulted in "New Poverty". The discussion will begin with a general discussion of poverty, particularly in the developing world because working street children are considered to be mainly a developing world problem, at least for the time being. What is "new" in poverty? What are the factors leading to such a transformation? The relevant factors will be organized around the global dynamics and the neo-liberal paradigm; subsequent changes in the market, state and family. The consequences of this process on children, particularly in its extreme forms that resulted in the emergence of working street children, will be analyzed next.

2.2 Poverty in General

Poverty, as an economic concept at first sight basically referring to a problem with income, is a quite intriguing topic to study. The history of poverty is as old as the history of humanity; however, even what it refers to is an ongoing debate. Who is poor in which society depends on the criteria used for definition. The fact that the definition of "poor" is frequently made by those who might not be considered to be poor further complicates the concept (Erder 2004, 40). Moreover, poverty is multi-faceted in terms of its characteristics, causes and consequences, and the way it is reflected in different social structures. Contrary to the first impression, poverty is beyond the economic dynamics. Its causes are related to the particular political paradigm globally and locally; the social values attached to certain sections of the society; factors such as age, race, ethnicity, gender and geography as well as the global process of the world economy.

To begin with, poverty in general and the definition of the "poor" might have political implications. The political choice to define "who is poor" interacts with the policies to be devised to eradicate or reduce poverty. Therefore, these policies might move certain groups out of poverty while drawing those excluded from the definition into poverty, no matter it does directly or indirectly. The social construction of the "poor" might result in a perception that the poor are lazy, criminal and different, even as inevitable and necessary determinant of macroeconomic prosperity; which might in turn result in their exclusion from the society (Becker 1997). Nevertheless, the contradicting understanding might lead to integrative policies.

On the other hand, poverty is shaped by the factors such as age (as the elderly and children are the largest risk groups) (Dziewiecka-Bokun 2000, 256), race and ethnicity (as people from the disadvantaged race/ethnic groups are more likely to suffer from poverty), gender (as existence of gender inequality leads to income inequality between male and female, particularly the women leading single-parent households are more likely to be poor) (Burchardt 2000, 390), and geography (as poverty varies from country to country, as well as in different region and societies) (Cornia 2001, 4-10).

Finally, poverty is affected by the changes in the global economy. It is exposed to influences from the international developments, therefore local definitions and policies might prove short of eradicating or reducing.

These multifaceted characteristics of the poverty are accompanied by the fact that poverty is a dynamic process. It refers to a certain state of living conditions of particular groups in a certain geographical unit for a certain time period; nevertheless, it is subject to constant change over time and space. Specifically speaking, the types of families and households at risk of

16

poverty vary from country to country over time; "poor" groups in a particular country might also change in time. An evaluation of dynamic nature of poverty and the discussion of recent transformation of poverty leads to questioning the newly added dimensions of poverty in comparison with the "old" or traditional aspects.

2.3 New Poverty and "Old" Poverty

The dynamic nature of poverty implies that poverty survives over ages, however, its form might change. Therefore, what is meant by *New Poverty* does not signify a sudden break, but rather it points out to certain changes and newly emerged aspects in a historical continuum. The transformation poverty is claimed to go through in 1980s refers to such a situation.

To begin with, poverty in a certain society used to be shaped more by local or national dynamics, and the effects of global instabilities or trends were comparably limited considering that the states were not as connected as today they are, economically, socially and politically. Factors such as natural disasters, concentration of land and natural resources in the hands of certain groups and local economic crises were more influential in driving people in and out of poverty (Nissanke and Thorbecke 2005, 3), rather than an economic crisis in Southeastern Asia. Today, these facts still play role on poverty. However, the effect of global dynamics increased dramatically. Explaining the causes of poverty in a certain society has more and more to do with the global political economy.[1] In other words, poverty transcended the national borders in 1980s in the way it has never done before (Nissanke and Thorbecke 2005).

Internationalization of poverty has come about in two ways: first of all, the international dynamics has become more and more influential on the rise or decline of poverty in different countries; and, secondly, countries with different political and socio-economic background have suffered similar reflections and forms of poverty. In other words, poverty has been produced by international factors and it is transferred to different geographies.

---

[1] Global political economy is used in this study in the sense of interactions between political and economic actors on the national and international level, not necessarily through the mediation of state mechanisms (UNDP 1999).Emergence of new actors such as international monetary organizations and multinational corporations, emergence of new global markets and integration of national markets through increased international trade and relations with the international organizations (IMF, World Bank and World Trade Organization especially), dominance of new rules and norms in line with neo-liberal paradigm and new tools of communication and transportation characterizes the global political economy.

In addition to increasingly global character of poverty; in the contemporary form of poverty, new dimensions accompanied the traditional problems. Here, two conceptual tools help to evaluate the difference. The first conceptual tool is the two-fold definition of poverty, absolute and relative, emphasizing the relative dimension of poverty in the contemporary form. *New Poverty* refers to deprivation likewise the poverty in the traditional sense, nevertheless, this deprivation is less to do with non-existence of opportunities and resources than access to the available opportunities. The second conceptual tool puts emphasis on social exclusion and inequality. This second conceptualization is related to the division between *absolute poverty* and *relative poverty* as relative poverty signifies the existence of inequality likely to stem from social exclusion. The term "social" in exclusion refers to the role of political, social and cultural aspects of exclusion in *New Poverty* although the economic exclusion has been more dominating in the traditional ways of deprivation (Gordon and Townsend 2000, 26). In the following parts, these two-fold conceptions of poverty; firstly absolute poverty and relative poverty, and then social exclusion and inequality will be discussed.

2.3.1 Absolute and Relative Poverty

There are various definitions of poverty developed in time by different scholars. In general terms, poverty might be understood as a concept in connection with low income or no income or as a consequence of insufficient income reflected in consumption patterns and living standards (Özcan 2003, 2). Recent studies tend to have a two-fold definition for poverty, generally cited as "absolute" and "relative" poverty. Although these definitions might be different in different resources, in general, absolute poverty measures the poverty in terms of personal conditions while relative poverty compares the living conditions of a certain person, or group, or nation with the rest. Precisely, absolute poverty is identified with "…severe deprivation of basic human needs, including food, safe drinking water, sanitation facilities, health, shelter, education and information. It depends not only on income but also on access to social services" (UN 1995, 41). The last sentence of this definition leads to the definition of relative poverty. Relative poverty implicates the existence of the conditions defined for absolute poverty and moves beyond by referring to not being able to benefit from or not having access to the services, goods and opportunities most people in that society or in another country take for granted (Gordon 2000,

49-50). This two-fold definition might also be employed in order to distinguish between the traditional forms of poverty and the *New Poverty*.

As frequently mentioned before, poverty is a dynamic process and *New Poverty* is a phase of the ongoing process which is also subject to change over time. Claiming that *New Poverty* is different from the traditional forms of poverty does not necessarily mean that the traditional causes such as low income, economic and political crises and natural disasters disappeared, but rather new risk profiles accompany the traditional factors. *New Poverty* refers to a condition in which both the international dynamics are more important (Townsend 1993) and poverty itself has become international in that poverty might be reflected in similar forms in totally different countries. It signifies a problem with low income, but also indicates that deep and increasingly permanent pockets of poverty exist even within rich nations with high economic growth and Gross National Product (Gordon 2000, 25). The most unique characteristic of *New Poverty* is that the expansion of economic growth goes hand in hand with the increase in inequality and poverty. In its new form, poverty manifests itself not only as lack of income, hunger, malnutrition, ill health, limited or lack of access to education and other basic services, increased mortality from illness; but also as increased homeless and inadequate housing, unsafe environments, social discrimination and exclusion. Moreover, lack of participation in decision-making, civil, social and cultural life is experienced in *New Poverty* although these opportunities are available for the rest of the society. The dynamics behind the differentiation of absolute poverty from relative poverty invokes the role of social exclusion and inequality in that why relative poverty have become pressing.

2.3.2 Social Exclusion and Inequality in the context of Poverty

New Poverty emerges as an interaction between social exclusion, inequality and poverty. The relation between poverty and social exclusion is closely associated with the relation between poverty and inequality. Inequality in the society leads to the fact that even the economic growth does not eradicate poverty since the advantages of the economy do not reach to those socially excluded and/or benefiting less from income distribution. If poverty traditionally refers to the lack of material resources, social exclusion refers to cultural and political deprivation in addition to economic exclusion. A person might not be poor in the sense of having somewhat regular

19

income to sustain his/her basic needs, but still might be excluded for the reasons related to race, ethnicity, gender, being disabled and age (Blakemore 2003, 80). In this sense, social exclusion for women might be claimed to be more serious since women might be exposed to social exclusion for race, ethnicity, physical condition and age reasons, however, women might also be excluded because of being women.

Hence, the two trends in relation to poverty-social exclusion nexus are simultaneous widening of the opportunities and improved living standards for a portion of the population and fragmentation of the poor into increasingly isolated and vulnerable groups. The problem lies in the fact that neither the benefits, nor the social costs of the changing global political economy are distributed equally in the society. In a sense, one's consumption might reduce the welfare of the other (Titmuss 1974, 60). The poorest sections of the society frequently shoulder majority of social costs through taxes and reduced welfare benefits. Social exclusion as a characteristic of *New Poverty* differentiates it from the traditional form of poverty in that poverty traditionally refers to mainly income disadvantage, lack of resources at the individual or household level, non-participation for lack of material resources while multiple disadvantages, community level resources and multiple causes for non-participation is valid for the *New Poverty* (Burchardt 2000, 387). Inequality also contributes to this distinction by further emphasizing the relative dimension of poverty. The persistence of inequality in the contemporary societies implies that a general improvement in the living standards does not eradicate poverty as all sections will not benefit equally (Holman 1978, 17).

The social risks are composed by class risks, life-course risks and intergenerational risks. Inequality is produced in the family through male-dominated family structure and state sponsored continuity of the model family (family benefits, family wage assigned to male breadwinner, discouraging labour market regulations for women by requiring uninterrupted careers for promotion and high income). Market further compounds the inequalities (Esping-Andersen 1999, 44). Since the families no longer absorb market failures, extreme poverty emerges resulting even in child labour.

Today, particularly refugees, immigrants, disabled and in general the women are socially excluded (Erder 2001, 58-63). Longer terms of unemployment brings economic exclusion, which in turn results in exclusion from cultural and political exclusion from the mental and material opportunities and resources available in the society (Buğra and Keyder 2003, 21). *New Poverty* with its nature makes it impossible to ensure social integration.

Until this part, it is aimed to explain why we refer to a new form of poverty and the newly emerged dimensions of poverty with the *New Poverty*. But why did such a transformation happen?

2.4 The Motives behind New Poverty

The multidimensional nature of poverty necessitates a multifaceted discussion on the process that led to poverty. In the case of *New Poverty* which is more susceptible to global forces as mentioned earlier, globalization is of particular concern. The effect of neo-liberal paradigm, the change in the market economy in the individual states and the overall change in the global political economy, transformation of state structures with decreased social expenditures and diminished intervening role in the economy and finally the change in the society and family form the basic tenets of the process that resulted in *New Poverty*.

Certainly, in different societies, some factors have been more influential depending on the local dynamics; however, it might be argued that all these factors more or less have been in interaction in most countries.

In the beginning of 1980s, two almost concomitant developments, particularly as a response to the previous problems in the world economy, have taken place: the rise of neo-liberal paradigm and globalization.

2.4.1 Neo-Liberal Paradigm

The first of these developments is the rise of neo-liberal paradigm not only as a world view, but also in the form of concrete reforms, particularly affecting the developing world in terms of poverty. The crises of 1970s resulted in the questioning of existing institutions and economic structure. At this point, neo-liberalism rose proposing remedies by changing the political-economic systems. In general terms, neo-liberalism claims for free functioning of the market with minimum state intervention or better without any intervention. Liberalization of trade with gradual termination of tariffs and quotas and privatization of public goods and services

21

are to be accompanied by cutbacks in the social expenditures together with withdrawal of subsidies and shrinking of the state in general. Deregulation of the labour market and flexibility of production are other premises neo-liberal paradigm rely on (Ecevit 1998, 23).

From the perspective of poverty, neo-liberal paradigm is influential as for it prevents redistribution of income through social policy provisions such as free health, education and housing assistance (Topal 2004, 13). Furthermore, it foresees unorganized labour force and tolerates informal jobs both of which lower living standards of people with lower wages and benefits, and less job security.

Neo-liberal paradigm affects poverty particularly in the developing countries through the structural adjustment programmes (SAPs) imposed upon developing countries with debts and aiming to be a part of the international system.

The initial emergence of the SAPs is to do with the aim of providing a framework in which in-debt countries would make the necessary reforms to improve their economic performance. The SAPs call for export-oriented production, decreased domestic demand, cutbacks in public spending, shrinking of state, deregulation, flexibility, disorganization of labour force and low labour cost. The SAPs as concrete economic measures designed in accordance with neo-liberal paradigm have been applied in 1980s. Under SAPs, the countries transferred their resources as part of the debts, therefore could not make the necessary investments for economic growth. Production diminished accordingly. The real wages and benefits declined both because of the increase of inflation and weakening of labour unions. High employment rates added to the withdrawal of previously free services (i.e. health and education) the living standards fell dramatically (Ecevit 1998, 34).

One should not consider the implications of neo-liberal paradigm in itself, but rather it should be analyzed in interaction with outpacing of globalization as the latter significant development of 1980s.

2.4.2 Globalization

Here, a parenthesis might be opened in order to avoid confusion: what is referred here as "globalization" or "global dynamics", should be distinguished from the global interactions before 1980s. It might be claimed that globalization traces back to the first geographic discoveries and colonization period, but the term beginning with 1980s refer to a distinct situation. In 1980s, together with the rise of neo-liberal paradigm calling for free operation of market and liberalization of economy over the world by opening the national borders for free flow of at least capital; globalization has become a dominant factor shaping national and international political economy.[2] In a way, the rise of neo-liberal paradigm has provided a theoretical framework in which global forces or globalization process would have increased roles in national economies and political decisions. Thus interaction of neo-liberal paradigm with the global forces led to emergence of a new era, integration of globalization and neo-liberalism in this process. The *1999 Human Development Report* summarizes what is new with globalization since 1980s as new markets, new actors, new rules and norms and new tools of communication (UNDP 1999, 30).

The markets are claimed to be new as global markets emerged and the national markets are increasingly globalized, new deregulated financial markets and global consumer markets are created. New actors acquired roles in that multinational corporations and international organizations such as IMF, World Bank and WTO have become highly influential on even national politics and economies. New rules and norms in line with neo-liberal paradigm are set and multilateral agreements increased. Finally, the technological innovations in communication and transportation such as Internet, cellular phones, computer facilities and rapid transportation vehicles further shrank the world.

Coming back to the interaction between neo-liberal paradigm and globalization; it might be argued that each fostered consolidation of the other one. Nonetheless, globalization has not only been driven by market forces or technology, it is also an outcome of the liberal political and economic paradigm. Globalization offers economic growth and development to competitive societies and speaks through a paradigm which claims that liberalization in individual countries would decrease domestic prices, transfer technology, increase economic growth and improve income distribution (Cornia 2000). Accordingly, policy measures are foreseen to be taken by the governments for trade and financial liberalization, privatization, deregulation and disorganization of the labour market and cutbacks in the public services; which in return would increase the competitiveness of national economies, reduce the fiscal burden of the governments and improve income distribution (Nissanke and Thorbecke 2005, 3). Therefore, globalization and the new

---

[2] In this study, the term "international political economy" refers to an interaction between politics and economy on the international, particularly in the last few decades.

dynamics of global economy forced the states to make market reforms, but it also presented a political paradigm in which governments would voluntarily modify their social policies for the greater good (Nissanke and Thorbecke 2005).

The technological dimension of global dynamics have made it possible for neo-liberalism to work and advance, while the advance of globalization let neo-liberalism to become concrete through the Structural Adjustment Programmes (SAPs). This interaction between neo-liberal paradigm and globalization might be claimed to cause a change in the economic structure.

2.4.3 The Changes in the Economic Structure

The outcomes of this process transformed the structure of the market economy. The driving force behind the market economy was industrial production until 1970s. In late 1970s and early 1980s, the transformation in the economy moved to a post-industrial type. Service sectors and informal sector referring to not necessarily illegal but rather unregistered labour have risen. The new economy necessitated removal of national barriers in front of the international trade on the one hand, and on the other hand, required flexibility and deregulation in the labour market. Gradually, more and more temporary and part-time jobs with little or even without any social security replaced full-time, registered labour and life-long occupations (Esping-Andersen 1999, 24). Trade liberalization and increasing competition in the world markets also put pressure on the wages in the developing economies not only by increasing wage inequality and unemployment, but also by causing a decrease in the real wages in these countries (ILO 2001, ILO Working Party on the Social Dimension of Globalization 2001). The Fordist hierarchical working model gave way to decentralization even in the remaining industrial settings. The companies engaged in international trade have become multinational in character, producing in one country frequently through sub-contracting and marketing in different countries; thus becoming actors in decision-making in several countries. In other words, the structure of the occupational setting is transformed worldwide in the form of post-industrial economy.

The technological innovations in production, transportation and communication contributed to this process. Technology and capital-intensive jobs have increasingly replaced labour-intensive occupations driving many people who fail to provide the qualifications and education required by new jobs into poverty. Technological diffusion and access to technology and information remained limited as technology and knowledge are powerful assets for the technology producing countries for economic and political influence on other countries. As a result, the advance in technology not only caused a divide between skilled-educated workers and those lacking, thus a within country inequality, but also increased inequality between the countries (Nissanke and Thorbecke 2005, 9).

With globalization, flow of capital and labour increased within and between countries, though the control over the unskilled labour migration remained strict. Migration of skilled labour from underdeveloped and developing countries to more developed countries and migration to the more developed regions within the countries further increased inequality.

As mentioned previously, the interaction between globalization and neo-liberal paradigm have implications on the economic structure. Nevertheless, neo-liberal paradigm is a multifaceted framework. In addition to the economic dimension, influence of this process might also be discussed in the political and social sphere which needs a discussion of the change in the social policies in general.

2.4.4 The Change in the Social Policies

The reflection of these changes is also seen on the social policies. Justified by national economic prosperity and competitiveness in the global market, countries have exercised liberalization and privatization leading to a further increase in unemployment. Some of the welfare services have been withdrawn, education and health costs increased and means-tested benefiting replaced universal coverage in many countries, all of which might be summarized under the structural adjustment reforms have resulted in diminishing of the welfare state and increased inequality in income distribution (Gordon 2000, 5).

The structural adjustment reforms of the governments stems from global economic paradigm as well as from the question posed by globalization on the states to use political and economic tools to intervene in the economy for social provisions. It is also a political choice on behalf of national governments in terms of the new political economy and the change in the settings of the traditional welfare state. The rigid labour market policies contradict to the new economic environment and high unemployment figures are witnessed. Many states are poorer than the previous periods due to the relatively worse economic performance and the financial overload on the government due to the high public expenditures. In many countries, the crises have become permanent. Even the crisis are overcome, the economic growth no longer means automatic increase in the job opportunities as the structure of the economy shifted towards the service sector with declining labour-intensive jobs and fewer occupations demanding higher qualifications and more experience.

In the context of the new political economy requiring higher skills and education, particularly the youth and women, as they are usually less experienced and sometimes having lower education, increasingly experience difficulties at reaching the opportunities and benefits of the market. So, youth and female poverty have emerged to be persisting aspects of *New Poverty* (Esping-Andersen 1999, 103). The Golden Age of capitalism was successful in absorbing them in the assembly line production, but technology and new dominant pattern of employment potentially exclude them.

The employment opportunities are fewer; moreover, many available jobs are ill-paid, usually temporary and part-time, through sub-contracting in the informal sector, lacking working security. Thus, working poor have increased in number. Increased female demand to participate in the labour force resulted in transferring the secondary jobs to the women and sometimes to the children, feminization of poverty and employment became apparent (Buğra and Keyder 2003, 28). Throughout the second half of the 20th century, the class distinctions comparably blurred and ideal of middle class society was achieved to an extent, nevertheless, upward class mobility remained limited and lagged afterwards with budgetary cutbacks and the aim of intergenerational upward mobility remained short of spreading over the entire society. With permanent unemployment and poverty, withdrawal of free mass education in changing degrees thus increasing educational costs, declined household income and early age child labour resulted in the inheritance of poverty from the parents to the children. Consequently, social exclusion and marginalization became persistent and permanent for the poorest social categories (Esping-Andersen 1999, 33).

Single parent households, large and crowded families, the unemployed, people with no education or low education, old people, the sick and disabled, those who lack access to assets, young people and children are the new risk groups (Dziewiecka-Bokun 2000, 256). Faced with economic difficulties and corresponding withdrawal of social provisions in the majority of the developing countries, the familial structure is also effected.

The global dynamics gave way to a new international division of labour. In this division, low-skill requiring labour-intensive sectors headed to the developing countries where labour costs are cheaper. As previously conceptualized under feminization of poverty, women are particularly affected from this process (Gündüz Hoşgör 2001, 120). This period also coincides with the mass migration from the rural areas to the cities in search for jobs in the developing countries. The migrant families in most of which men are unemployed as they lack the necessary skills and education the new economy requires; women and children have become cheap labour supplies. Most of these women and children are employed through subcontracting in low-paid, part-time or temporary occupations. The relative cost of children when they are not working increased accordingly as the cost of education increased while the incomes are decreasing (IPEC 2006, 5). Some of the children who could not find jobs even in subcontracting and their families started to look for other options when streets emerged as new working places.

Emergence of working street children as one of the extreme reflections of the change in child poverty and poverty in developing countries might be discussed in this context.

2.5 Working Street Children, Child Poverty and New Poverty

Low income is one of the most important causes behind child poverty. Families with low incomes can sustain fewer of their basic needs which have a direct influence on the expenditures devoted to children (UNICEF Innocenti Research Center 2000, 12). Moreover, recently increased income inequality has caused a divide between the rich and the poor families, and the families with children suffered more from income inequality as the state transfers for them eroded in time (Daniel and Ivatts 1998, 226). Despite economic difficulties, the developing countries continue to have high fertility rates, low age of first marriage, low mean age of child-bearing, high birth rate outside marriage and large families (Bradshaw and Barnes 1999). In such unstable family conditions, the familial support to the well-being of children diminished, and children is suffering extreme levels of poverty even in the developed countries.

However, income or demography cannot fully explain re-emergence of child poverty. They are significant dynamics behind child poverty, but then how the variation of child poverty rates in countries with similar levels of economic development and demographic transformation

27

can be explained? The answer to this question leads to a third, and probably the most important factor: governmental policies and the political choice about children.

The factors behind child poverty as low income, income inequality, demographic transformation and the decline of available welfare incentives indeed show that the current situation of children is affected by the factors that led to the emergence of *New Poverty* in general.

Regarding the definitions of poverty as absolute and relative poverty, child poverty might be affiliated with both of them historically. As absolute poverty refers to lack of material deprivation from basic needs, children of the poor families also suffer from absolute poverty (Bradbury and Jännti 2001, 13). However, the contemporary form of child poverty is much more related to the relative poverty their families suffer from, as poverty of the families also transformed in time to a new form stemming from inequality and social exclusion (Bradshaw 2000, 237-239). Moreover, child poverty is also related to the relative poverty within the household as household income is distributed unequally in the families frequently privileging the adult members. The concept of relative child poverty might indeed be considered to be an indicator of relative poverty and inequality in a society. Identification of the United States among the countries with highest child poverty despite high economic growth and higher income level points to this claim that income inequality might be more affective on the child poverty rather than the general income level of a country (UNICEF Innocenti Research Center 2000, 2). Therefore, poverty, which the children suffer from contemporarily, might be explained with the dynamics behind *New Poverty*. In addition to the shared dynamics, child poverty is also affiliated with *New Poverty* in terms of the newly emerged aspects in the recent decades.

As a consequence of intergeneralization of poverty, child poverty contemporarily tends to be more permanent in the future of the poor children. The declining social mobility is likely to produce a cycle for the children of poor families in that the opportunities available for the children are fewer compared to the richer families. Higher costs of education as the government supports for education is diminishing worldwide and free education remains limited, the chance for the poor children to have a similar education with those of richer families is relatively low. They are more likely to suffer unemployment, to be lowly paid when they are working and to have health difficulties when they grow up (Vleminckx and Smeeding 2001).

In terms of gender emphasis, child poverty is today more affiliated with gender dynamics compared with previous periods. The discrepancy in the income distribution between men and women tend to increase child poverty in single-parent households most of which are headed by females (Christopher et al. 2001, 212-214).

Besides gender and inequality dynamics, the contemporary form of child poverty is much more to do with social exclusion for both the parental impoverishment is more related to the non-economic factors of belonging to disadvantaged communities of colour, ethnicity and religion.

The distinguishing circumstances of child poverty today also show itself in the way it is practiced. The children of the poor families have been working in household occupations; and in factories and workshops after industrialization. The economic value of children had dominated the poor families traditionally; however, the support provided by the welfare regimes prioritized the psychological value. Contemporarily, there seems to be a reverse process in this regard. First of all, with the impoverishment and declining incomes, families became less able of caring their children. The demographic transition in which parents are either dual workers so that the parental care has to diminish to sustain economic well-being or single-headed households increased meaning lower income and less psychological support. With the withdrawal of welfare incentives and consequent increase in education and health costs of children, the relative cost of children's not working increased more. Under these pressures, the contemporary poor children are no longer passive victims of poverty, but have to become active agents in the family survival strategies against the difficult living conditions. Previously, families somehow managed to make their children work at the available times remaining from school as a contribution to family budgets. The jobs the children were engaged with included selling small items, but they were more likely to work in a workshop or factory both for money and skill-investment for the future. In the initial phases of industrialization, the children were employed as cheap labour with nimble fingers (Goldson et al. 2002, 89), however, with the advance of technology requiring the skills and education children lack, and diminishing of manual jobs requiring physical work of children, the occupational settings for the children of poor families are also transformed.

This aspect of contemporary form of child poverty might be claimed to be the most direct consequence of the new conditions of poverty on children and the most striking change in the way poverty is displayed. The children are no longer working at the times out of school. Many of them are dropping the school in order to work (Akşit et al. 2001, xi). They are not only contributing to the family budgets, most of them are the sole breadwinners in their families (Kahveci et al. 1996, 38). The jobs they are employed are no longer skill-investing jobs as they are not much available for children as those kinds of jobs are vanishing in post-industrial economy and the remaining jobs are requiring the skills and education that the children cannot provide. Newly emerging jobs are usually hazardous for mental and physical development (Akşit et al. 2001, xi), even involving drug dealing and prostitution (Küntay and Erginsoy 2005, 53; Alexandrescu 2002, 7). It is for sure that such occupations have been historically abusing child labour; but what is new is that these jobs became widespread and increased in number therefore it has become more visible.

In this context, emergence of working street children in many developed (e.g. US and UK) and developing countries (e.g. countries in the Eastern Europe, Middle East and North Africa) constitute a striking example of the new form poverty and child poverty acquired. Globalization process and the new political economy gave way to unemployment in certain sections of society lacking necessary qualifications and increased poverty of these children's families but also directed the children to work on the streets. Other than income insufficiency, factors such as social exclusion and inequality and, of course, decreased state support to children and families with children played role in this process. As both the familial support and governmental support behind the children are withdrawn, the children in the *New Poverty* circumstances began working on the street doing various jobs. Some of them even began to live on the street almost losing their contact with their families. Continuing the discussion between working street children in the context of *New Poverty*, thus, needs clarification of what is meant by working street children.

2.6 Terminology on Working Street Children

The term working street children seems to be an ambiguous reference at first sight. As International Programme on the Elimination of Child Labour (IPEC), a sub-branch of International Labour Organization (ILO), is accepted as the internationally authoritative political body defining the strategies and perspectives on child labour, the terminology discussion in this chapter is based on relevant articles of ILO.

30

To begin with, child is defined as the individual under the age of 18 years (ILO Convention on the Worst Forms of Child Labour 1992, No. 182). Work is defined, on the other hand, in terms of economic activity and "covers all market production (paid work) and certain types of non-market production (unpaid work), including production of goods for own use" (ILO 2002, 30). Under the light of these definitions, working of children above 12 years in light work in a developing country is acceptable (ILO Minimum Age Convention 1973, No. 138 Article 3). Every country defines its own minimum age for work; however, it is generally defined as 15 years in global estimates (ILO 2002, 30). 12 years is the boundary age for light work, in which light work is defined as the type of work that is not harmful to health, development and does not produce prejudice to the child in her/his social environment. It should also not exceed 14 hours per week (ILO Minimum Age Convention 1973, No. 138). In this context, child labour covers all the children under 15 years age, excludes those are smaller than 5 years, and those who are between 12-14 years who work less then 14 hours per week unless their work is hazardous. The children of age between 15-17 in the worst forms of child labour are also included (ILO 2002, 32). Hazardous work is work, which is "likely to jeopardize or harm the health, safety or morals of children" (ILO Convention on the Worst Forms of Child Labour 1992, No. 182). The Unconditional Worst Forms of Child Labour is slavery, trafficking, debt bondage, serfdom, compulsory of forced labour, armed conflict, prostitution, pornography and illicit activities like drug production or trafficking (ILO Convention on the Worst Forms of Child Labour 1992, No. 182 Article 3). The final concept is children at work in economic activity. This concept is the broadest definition regarding working of children and covers also the illegal, unpaid, casual and informal work (ILO 2002, 15).

These children may be working in either visible or invisible jobs, therefore to maintain a definite statistical approach is not much possible. Hence, the data presented by the national and international bodies are more of estimation rather than fact sheets, and simplifications and exclusions had to be made for the sake of feasibility. In the year 2000, IPEC estimates the existence of almost 350 million children working worldwide. 211 million of this number is at work in economic activity, 190 million is considered to be child labour, 112 million in hazardous work and around 9 million children are in the worst forms of the child labour (ILO 2002).

Focusing specifically on working street children generally involves those children working on the street regardless of whether they spend the night at home or on the street. In earlier studies, a three-fold categorization is seen. This categorization is developed by UNICEF in 1986 on the children working and/or living on the street in the developing countries (Karatay et al. 2000b, 455-456). Basing on the relations with the families and the place where the night is spent, a categorization is made as "children with a continuous family contact", "children with occasional family contact" and "children without family contact (Atauz 1990a, 7). Atauz also refers to a popular reference to the children living on the street as *köprüaltı çocukları" (bridge-bottom children)* (Atauz 1990b, 9). More recently, terminology on working street children is developed on a distinction based on whether the children are permanently living on the street or not. In this categorization, the children are referred as "children working on the street" and "children living on the street" ("children of streets"/"street children") (Akşit, Karancı and Gündüz Hoşgör 2001, x). Children working on the streets spend the day, and even sometimes majority of the night on the street, but finally return home to their families. Although this relation between children and their families may not be considered as stable and absolutely regular, it is possible to claim that their connection is going on and somehow the children are protected and supervised by their families. The second category of the children has broken their ties with families or families themselves disintegrated. Some scholars refer to these children as homeless children (Fazlıoğlu 2002, 66). There are also other studies making a classification of working street children on the bases of tasks of earning (Altıntaş 2003, 12). In this study, children living on the street and children working on the street will be used.

In the following chapters, working street children in Turkey and Romania will be discussed within the framework of this theoretical context as the internationalized new form of poverty made two such different countries with distinct political, economic and social background suffer from the same example of new child poverty: working street children.

CHAPTER III

WORKING STREET CHILDREN IN TURKEY

3.1 Introduction

In the previous chapters, it has been stated that poverty has transformed and acquired a new form referred as *New Poverty* identified with persistence of relative poverty, inequality and social exclusion. Furthermore, it has been mentioned that child poverty has transformed in line with the transformation of poverty in general, overwhelmingly under the influence of global dynamics. The previous chapter is concluded with emergence of working street children as a new form of child labour in these circumstances; and a distinction between "children working on the street" and "children living on the street" or "children of streets" (popularly referred as "street children") is made for a better understanding of what is focused in this study.

This chapter begins with a review of the existing literature on child labour and working street children in Turkey. Next, child poverty in Turkey after 1980s is discussed, particular emphasis put on the factors leading to the increase and transformation of child poverty in Turkey in the last 25-30 years. Emergence of working street children as a form of child poverty and as a worst form of child labour follows the discussion on child poverty. The scope and dimensions of working street children is considered finally together with the response measures taken against the problem.

3.2 Review of Literature on Child Labour and Working Street Children in Turkey

In terms of the literature on child labour and working street children in Turkey, there are studies particularly since 1980s. Among earlier studies, there are research by Balamir (1982), Atauz (1989a; 1989b), Yener and Kocaman (1989), Konanç (1991) and Ozbay (1991). Child labour and child population are the dominant themes in these studies, however, study of Konanç (1991) focuses on the working street children in Ankara.

The studies of Atauz (1990a, 1990b, 1995, 1996, 1997), Ertürk (1994), Köksal and Lordoğlu (1993), Konanç (1996), Kahveci et al. (1996), Özbay (1998), Oto et al. (n.d.), Küntay et al. (1998), Küntay and Erginsoy (2000; 2005), Karatay et al. (2000a, 2000b) and Altıntaş (2003) are the main studies on child labour after 1990. Among these studies after 1992, the study of Atauz (1997) is about working street children in Diyarbakır.

"Child Labour and Street Children in Turkey" by Sevil Atauz (1989a) is the among the first field research on street children in Turkey. In this study, Sevil Atauz gives background information on child labour in the world dating back to 19th century industrial settings. As child and woman labour are cheaper, historically child labour exist, especially in the rural areas. However, working of children in the cities of the developing countries is more recent and it is an outcome of rural to urban migration (Atauz 1990b, 1-3). In terms of children working on the street, the study is mainly on street children. Street children are defined as "all those children without families, runaways, juvenile delinquents, abandoned children, children in need of care and protection, child labourers, maladjusted children" (Atauz 1990b, 35). In this study, in order to define the children on the street, she uses a categorization which is also used in Atauz's later studies. Children are categorized as "children with a continuous family contact", "children with occasional family contact" and "children without family contact" in this study (Atauz 1989a, 35). It is stated that street children only compose one third of all children on the street and most of the children still have contact with their families. By the date of the study, street children are mentioned to be visible in İstanbul, Ankara, İzmir, Adana, Diyarbakır, Antalya and Gaziantep. Age of street children might be as low as 4 or 5 years. With the exception of gypsies and beggars, the study mentions that almost all of the street children are boys (Atauz 1989a, 35-37). Increased economic difficulties, the change in the family structure with migration, psychological factors (desire to have autonomy from their families) and education (increased cost and decreased quality of education) are stated as the factors that lead to emergence of street children in Turkey (Atauz 1989a, 39-42).

Sevil Atauz's research in Ankara and Şanlıurfa on street children (1990a) is important in that this study also refers to the period before 1990. Atauz discusses rural to urban migration beginning with 1950s in Turkey. She relates migration flows in this period to technical modernization in agriculture and commoditization and consequent transformation of the rural structures (Atauz 1990a, 5). The scarce resources in the cities available for the new migrants caused deterioration in the living conditions which required child labour as part of the family survival strategies. Street children among these children constitute the main topic of the study. Atauz states that street children exist since 1950s especially in İstanbul, popularly known as *köprüaltı çocukları* (bridge-bottom children). Mothers or fathers of these children are usually in prison, a finding which refers to family disorganization as a factor behind emergence of street children. Street children living in İstanbul, Ankara, İzmir, Adana, Diyarbakır, Gaziantep, Şanlıurfa and Antalya are mentioned in this study, paying specific attention to the street children in Ankara and Şanlıurfa. According to this study, there were around 9000 street children in Ankara in 1990. The children are between 6-17 years old, most of them being 11-14 years. According to the research in Ankara, there are a few girls among street children and girls are usually engaged in selling and are working in more "passive" jobs (Atauz 1990a, 13). Most of the children completed elementary school, but very few are still going to school. Families of street children in Ankara are from rural origin and they are disintegrated. Death, divorce and violence are identified among the causes of family disorganization. In general, education level of parents is low, and literacy rates are particularly low for mothers. Unemployment rate is high among fathers. Although Atauz focuses on street children in this study, she also refers to children working on the street and having connections with their families. In this study, again, the author differentiates between "children with a continuous family contact", "children with occasional family contact" and "children without family contact". Family disorganization rate increases respectively, in line with this categorization. Children from the first group are engaged in selling petty commodities. Children in the second group contact with their families from time to time and live in houses for single people. They tend to gather with other children from the same origin city and work in similar jobs. Finally, street children do not work. This situation indeed signifies importance of family support for even working on the street doing jobs which do not require much skill or investment (Atauz 1990a, 15). A striking finding of the study on Ankara is that drug and alcohol usage rate also correlates with the groups. Some children are raped and homosexual intercourse are claimed to be common among these children (Atauz 1990a, 27). Criminal activity exists in all groups, however, in the first and second group, crime rate is lower and limited to theft usually. Atauz states that organizational crime is widespread in the third group.

In the research on Şanlıurfa, more recent migrants, those came to the city after 1980s are emphasized. In this city, children on the street are aged between 5-17, most of them being 11-12 years old. The children in Şanlıurfa are slightly younger than those in Ankara; as working on the street, usually selling petty commodities, are subject to negative reaction if the child is older. It is also stated that no girls are found to be living on the street in Şanlıurfa (Atauz 1990a, 28). Majority of the children completed elementary school; however, number of children who have never attended or left school is higher. Number of siblings in the families is higher - around 4-5- and family disorganization is again a factor for living on the street. Scarce job opportunities in the city are also reflected in the case of children. Only those children in the first group can be claimed to work by selling commodities like water, religious books and fish. Half of the children in the second group and most of those in the third group do not work. As the study focuses on street children specifically, family dimension is emphasized in Atauz's research. Atauz have another study, this time focusing on street children only in Ankara in the same year. The findings of this research are in line with her research mentioned above (Atauz 1990b).

Kahveci et al. (1996) covers shoe-shine boys in İzmir. The study is conducted by in-depth interviews with 100 boys. It is emphasized that shoe-shine boys in İzmir are children working on the street, not living on the street and they have continuous contact with their families. It is also stressed that these children are not working for pocket money, but they are the only breadwinners in the households. According to the study, shoe shine boys are aged between 6 and 15 years, and their average is 12. An important portion of them have never attended school, and most of the rest have dropped out after elementary school. In terms of the causes, the study identifies migration as the principal dynamic, a finding which is consistent with other studies conducted for instance in İstanbul (Karatay et al. 2000a; 2000b), Ankara (Altıntaş 2003), Diyarbakır (Akşit et al. 2001), Şanlıurfa (Atauz 1990a) and Adana (Akşit et al. 2001). The study states that "...the more recent their arrival, the poorer they are..." (Kahveci et al. 1996, 40). Families of shoe-shine boys are from rural origin, most of them migrated after 1980, on the average they have 6 children, fathers are either unemployed or underemployed, and particularly those coming from Eastern Anatolia are from Kurdish origin (Kahveci et al. 1996, 41). The authors express that shoe-shine boys are part of the broader range of child labour pointing to the fact that the children mentioned other children in the neighbourhoods working in different tasks. Majority of them were previously engaged in other jobs before shoe-shining (Kahveci et al. 1996,

47). Some of the children mention that they were exposed to sexual abuse (Kahveci et al. 1996, 57). This finding of Kahveci et al. on boys, combined with findings of Küntay and Erginsoy (2005) on girls might be assessed as incidence of sexual abuse in the case of working street children.

Küntay and Erginsoy's study (2000) is particularly dealing with teenage sex workers, which is a quite rare field of study in Turkey. The recent study by Küntay and Erginsoy (2005) specifically focuses on the girls engaged in prostitution in İstanbul. The study relies on a long term research dating back to even 1960s by Küntay in general, however, it specifically derives information from a study completed in 1998. In terms of working street children, Küntay and Erginsoy's study emphasize that with emergence of working street children, sexual exploitation of children has acquired a new dimension. The study focuses on the girls; however the authors acknowledge that boys are also under the risk. Main reasons for the girls to end up on the street are identified basically on the factors relevant to family structure. In this context, abuse and violence in the family and forcing the girls marry frequently in order to be paid by the groom's family (*başlık parası*) are identified. Besides, extreme value attached to virginity is mentioned as a factor. These events are claimed to cause trauma and "street" emerges as a solution to the pressures of family environment (Küntay and Erginsoy 2005, 16). Rural to urban migration, increased economic difficulties and commoditization of sexual relation in the great cities are identified among the other factors. In terms of the number of children engaged in prostitution, the study emphasizes the difficulty to reach exact figures; however, the number is claimed to be well below the other countries, for instance in the Far East Asia. Parents of the girls are in majority low educated, unemployed or working in informal sector for low wages. Family disorganization due to violence, death and alcohol/drug addiction is common among these families being as high as 75%. Girls were either not working, or working in textile industry, hairdressers and small shops previously. Girls meet/find customers in bars, hairdressers, restaurants, houses of the mediating people, poolrooms, bridges, gas stations and along the highways. Education level is also low among the girls due to the factors that their families might not want them to go to school for cultural reasons or economic difficulties or girls were reluctant to go to school due to mistreatment of teachers, poor performance or a desire to go to the entertainment places as an attempt to be free from the pressures at home.

Karatay et al. (2000a) focuses on the working street children in Beyoğlu district in İstanbul, and in another study of the same authors (2000b) İstanbul is focused in general in terms of working street children.

In the study on working street children in Beyoğlu district (2000a), the study emphasizes that the problem of emergence of working street children is a "global" issue (Karatay et al. 2000a, 425). This study criticizes three-fold categorization on working street children (which is also used by Atauz 1989a, 1990a, 1990b, 1996) and employs a more detailed approach as "totally abandoned children", "children of the street", "children on the street" and "children living with their families". Number of working street children in Turkey being lower than other countries with similar economic problems for instance in Latin America is explained by stronger family structure (Karatay et al. 2000a, 432). Main factor for emergence of working street children in Turkey is referred as "extraordinary migration" (*olağandışı göç*). By extraordinary migration, sudden and forced migration is meant. This study also emphasizes that families of the children have come to İstanbul 10-15 years ago. Terror and unemployment are cited as the main dynamics behind migration of the families. One third of the mothers cannot speak Turkish (Karatay et al. 2000a, 442). Families are crowded, most of them having at least 3 children. Majority of the fathers are employed, but they have low income. Half of the households have at least one sick person which might be caused by poverty. Around one third of the children are girls, and they are usually engaged in selling small items. The study identifies a relation between age and task of earning as the younger children are engaged in selling small items more than the older ones (Karatay et al. 2000a, 447). Contrary to the other studies (Kahveci et al. 1996; Atauz 1996, Akşit et al. 2001), majority of the children are going to school. It is also observed in the study that majority of the children working on the street in Beyoğlu district is monitored by their families and relatives. The study, in general, has a positive attitude to the problem in that it concludes that problem of working street children in Turkey, judging by the situation in Beyoğlu district, is not yet a problem with no solution unlike the countries with similar development levels.

In the other study focusing on the working street children in İstanbul in general, Karatay et al. uses an approach of examining the issue within the greater social context the children are living in. Street children are paid specific attention in the study (Karatay et al. 2000b). Average age of children working on the streets in İstanbul is around 12 years. This finding is higher than the average age in Beyoğlu (Karatay et al. 2000b, 474). Over 90% of the children are boys, girls have a lower average age compared with the boys, majority of the children have at least 3 brothers/sisters, and almost half of the children are either never attended or dropped out of elementary school. Shoe-shining is the most common task of earning followed by selling petty commodities. Almost 90% of the mothers are not working, and over 70% of them are illiterate, fathers are unemployed or working for low wages (Karatay et al. 2000b, 510-513). The study emphasizes high rate of shoe-shining, families being recent migrants from Eastern and Southeastern Anatolia, one fifth of the children migrating without their families and living in houses for singles, and grouping in tasks in accordance with origin city (a parallel finding with the study of Altıntaş (2003) in Ankara). This study argues that family disorganization is not a significant factor unlike other countries.

Altıntaş (2003) conducts research on working street children in Ankara making a classification on the basis of task types and emphasizes the role of being from the same hometown (*hemşehri*) in determining the type of work on the street.

In this study, data on child labour mainly depends on *Child Labour Survey 1999* (State Institute of Statistics 1999). Atauz (1989a, 1990a, 1990b, 1996), Kahveci et al. (1996), Karatay et al. (2000a, 2000b), Küntay and Erginsoy (2005), Akşit et al. (2001), Gündüz Hoşgör (2004), Gündüz Hoşgör et al. (2005), and Dayıoğlu and Gündüz Hoşgör (2004) are among the main research on working street children in Turkey. This study is mainly based on these research activities on Turkey, both because these resources provide objective and systematic data and also methodological opportunity for comparison with the research on Romania.

3.3 Child Poverty in Turkey

Depending on the close relation between poverty and child poverty, it is to be noted that scope of poverty in Turkey matters in the context of child poverty. To start with, according to the latest *Human Development Report* released by UNDP, Turkey ranks 94, in terms of life expectancy, educational attainment and adjusted real income (UNDP 2005). These ranks acquire meaning when it is compared to countries experienced severe economic crisis almost concurrently with Turkey such as Argentina ranking 34, and countries of former Eastern Bloc which had economically difficult transition stages such as Bulgaria at 55, Czech Republic at 31 and Romania ranking 64. The rank of Turkey rises to 70 in terms of GDP per capita at purchasing parity, which means that Turkey does not achieve the human development level that its economic development might allow if used properly (Buğra and Keyder 2005, 7). In terms of the nature of poverty in Turkey, extreme poverty (percentage of people living on less than $ 1-a-day) is at the level of 0.03% of total population, while the level of extreme poverty reaches to 1.8% on the basis of consumption and 3% on the basis of income (Buğra and Keyder 2005, 7). On the other hand, rates considering the level of relative poverty in Turkey are more striking. Relative poverty, measured by less than 60% of the median income, is 23% in Turkey. Gini coefficient of Turkey is 0.46 for Turkey which means that Turkey has the highest inequality rate in terms of income distribution among the members of EU and among the candidate countries (Buğra and Keyder 2005, 20; Eurostad 2004).

The reflection of this situation on child poverty is seen clearly at the figures. The report released by UNICEF annually on the situation of children worldwide contains basic indicators for child poverty (UNICEF 2004). Turkey ranks at 79th country among almost 200 countries worldwide in terms of under-five mortality rate measured as probability of dying between birth and exactly five years of age expressed per 1000 live births, value being 39 (UNICEF 2004, 105). In terms of nutrition indicators, in Turkey, 16% of infants with low weight which is equal to the world percentage including figures for even Africa (UNICEF 2004, 113). Literacy rates for Turkey are comparably higher compared with the mentioned indicators, figured as 93% for males and 77% for females, however, despite huge efforts during the Republican period, it can be argued that the figures are hardly above certain African countries. Finally in terms of child marriage measured as percentage of women 20-24 years of age married or in union before 18, is 23% between 1986 and 2003 (UNICEF 2004, 139). Overall considered, child poverty rate in Turkey is close to 20% according to 2000 figures (UNICEF Innocenti Research Center 2000, 17); ranking Turkey among the worst performing countries in OECD.

The picture gets much clearer considering Millenium Development Goals (MDGs) for Turkey. According to the MDGs Report for Turkey released in 2005, gender gap in primary education is closing, child mortality rates are decreasing, progress is achieved in terms of child diseases; however, poverty, gender inequality, child mortality and maternal health remain to be posing serious threats to child well-being in Turkey (UNICEF Turkey 2005). The report points to the fact that child poverty rate is well below the general poverty rate in the country. The threat of poverty for children in Turkey means that thousands of children die every year, 1 million children of elementary school age do not attend school, over 1.6 million children are engaged in child labour and number of children living and/or working on the streets is increasing despite efforts (UNICEF Turkey 2005, 5).

Child poverty in contemporary Turkey is closely associated with the rise of New Poverty distinguished by inequality, relative poverty and social exclusion. The report providing ranks for countries in terms of under-five mortality rate also refers to the share of household income between 1992 and 2002 stating that lowest 40% gets 17%, while highest 20% gets 47% (UNICEF 2004, 109). This relation implies that child poverty and low standards of children in Turkey is related to inequality. According to a study conducted by Ercan Dansuk based on State Institute of Statistics 1987 Household Income and Consumption Survey, absolute poverty declined from 32,01% in 1973 to 15,16% in 1987 (Dansuk 1997, 49). As stated above, relative poverty increased, considering together with high child poverty, the role of relative poverty on child poverty might be claimed to be meaningful. Furthermore, there is extreme polarization in the urban space particularly in metropolitan cities (Buğra and Keyder 2005, 15). Socially excluded families and their children are deprived of available resources and opportunities in the society (Micklewright 2002, 12), which might also be claimed to increase child poverty in Turkey leading to extreme forms of child labour as family survival strategies that is to be discussed in the part on working street children.

3.3.1 Factors Leading to Increase and Transformation in Child Poverty in Turkey

Child poverty in contemporary Turkey is associated with New Poverty in the sense that child poverty is also subject to transformation so does poverty in general. Therefore, it might be argued that factors that have led to an increase in poverty, particularly relative poverty worldwide, are also influential on Turkey. These factors are more relevant to global dynamics particularly influential since 1980s on the global political economy and accompanying rise of neo-liberal paradigm. Briefly, structural adjustment programmes, the change in the economic structure and social policy provisions, divide generated by structural transformation in the economy and technological innovations and the rise of consumption culture might be considered under this heading.

41

Nevertheless, it is to be noted that the current scope and trend of child poverty in Turkey has sui generic or country-specific determinants. Demographic factors, migration patterns, the 1999 earthquake comprise these kinds of factors in Turkey.

3.3.1.1 Global Dynamics and Neo-Liberal Paradigm

Turkey's industrialization strategy was import-oriented before 1980s in the sense of government control in the economy and labour market and relatively comprehensive social policy provisions (*Turkey Demographic and Health Survey 2003*, 5). In 1970s, Turkey experienced a balance of payments crisis. The import-oriented industrialization strategy with protectionist foreign trade policy, subsidization had caused difficulties in foreign payments. Combined with the decrease of currency reserve and subsequent in-debtness to the international monetary crisis after the 1970 Oil Crises; the crisis led to questioning of the policies (Ecevit 1998, 34). In this period, with increased debts to the international monetary organizations, Turkey is incorporated into the structural adjustment programmes (Boratav et al. 2000, 3). Initially, import-substitute industrialization is replaced with an export-oriented strategy with January 24, 1980 decisions. The aim has been to increase exports through increasing production by decreasing wages and salaries, weakening labour unions in order to prevent suspension in the production process and overcome wage increase demands; and increasing prices in order to control the domestic demand (Ecevit 1998, 35). Between 1981 and 1988, commodity trade liberalization is used as the main tool in order to integrate to the global market (Boratav et al. 2000, 5). Trade liberalization continued in the following periods, and privatization policies are applied and public sector employment decreased accordingly (Boratav et al. 2000, 22). Partly under the influence of global economic dynamics of structural adjustment programmes, and partly within the neo-liberal paradigm dominant in the world proposing these requirements, these measures are applied and decreasing role of state also meant declining social provisions available in education, housing, health and labour market. The social repercussions of these policies are reflected as distortion in income distribution (Ecevit 1998, 36), inflation, unemployment, loss of jobs in public sector, and worsening of wages for working people (which increased number of working poor). Average annual poverty rate increased by 4% between 1987 and1994 (Erman 2003, 42).

42

April 5, 1994 decisions further worsened the situation. Particularly after November 2000 and February 2001 crises, poverty has become a severe problem. Job creation has remained limited in Turkey. Between 1980 and 2004, population increased by 23 million, however, around 6 million jobs are created. Despite 65% labour participation rate among EU-15 countries, in Turkey, labour participation rate remained at 48.7% dragging Turkey among the countries with lowest labour participation rates in the world (World Bank 2006c, ii). Percentage of working poor increased to 50% of working population after these crises (Erman 2003, 43). In these circumstances, cutbacks in social expenditures have deepened poverty. On the other hand, Turkish economy went under structural transformation likewise other parts of the world. Transition from industrial production to a post-Fordist and post-industrial economy worldwide with the rise of service sectors has found correspondence in Turkey. However, Turkey has not gone through the regular stages that the developed economies passed through in that those countries had switched from agricultural production to industry and then to service sector economy. This transition is more of a leap attempt from agriculture to service sectors meaning that many new migrants from rural areas lacking the necessary skills and educations increase with technological innovations worldwide. As a result, structural transformation in the economy could not absorb Turkish workforce (Buğra and Keyder 2005, 13). Participation in the labour force (for all people over 15 years), which tends to decline since 1960s, have further decreased, from 71% in 1980 to 48.7% in 2004 (World Bank 2006c, 9, 36). Prevailing occupations are mostly informal sector part-time, temporary jobs frequently lacking social security, mostly subcontracting home made jobs.

These developments might be interpreted under three headings: first of all, poverty increased with decreasing incomes. However, this is rather an increase in relative poverty, as well-skilled and highly educated people have found the opportunity to generate considerable wealth, while those lacking these requirements are either unemployed or underemployed (Buğra and Keyder 2005, 17). Economic polarization has a spatial dimension especially in the metropolitan cities, scattering the rich to the sea costs if available and the poor to the transitory areas (Erman 2003, 43). This finding leads us to the fact that inequality is accompanied by social exclusion in the cities of Turkey.

Combined with decreasing state support to families and children, transformation of economic structure providing fewer jobs for lowly skilled and lowly education people, increased

43

unemployment and inflation; Işık and Pınarcıoğlu state that, particularly after 2001 crisis, poverty is transformed in Turkey from a form of poverty which is, at least to an extent, manageable through community networks and squatter settings, to another form which is not as manageable as it was before, because the available networks and housing is not available anymore (Işık and Pınarcıoğlu 2003, 51-53).

Repercussions of this process of globalization and neo-liberalism on the economy providing fewer jobs for families and children, increasing unemployment and inflation; on the politics decreasing state support through decreased social policies and on the family decreasing familial support through increased pressures on the family to involve children in the survival strategies; might be claimed to transform the way children suffer from poverty. With increased family and child poverty and transformation of poverty to a more hardly manageable form for the families; implications of the global dynamics have led to a process which created streets as an option for the children of poorest families.

3.3.1.2 Country-specific Factors in Child Poverty in Turkey

Contemporary situation of poverty and child poverty in Turkey cannot be fully understood without specific conditions in Turkey. Mentioning these factors for Turkey, does not mean that they are irrelevant to international circumstances, but rather points to the argument that although other countries might experience similar situations, the dynamics are more relevant to specific conditions of Turkey. While 2000 and 2001 economic crises might be considered to have international determinants, for instance the devastating earthquakes in Turkey in 1999 in Istanbul and Düzce, pushed many people into poverty since they lost their property and jobs.

Migration is an important factor in Turkey in the terms of poverty and the way children are affected from poverty. Rural-to-urban migration in Turkey has begun in 1950s after modernization in agriculture and declining opportunities in the countryside. However, these waves of migration have chain pattern, firstcomers to the city providing assistance networks to the followers through closely knit community ties (Akşit et al. 2001, 14). The most important instrument for integrating to the city had been building squatter houses, known as *gecekondu* (Buğra and Keyder 2003, 22). In time, they could add floors and rent them, many of them have been legalized during the election campaigns of 1980s. Today, it is not as easy as before to build a *gecekondu* as land is much more precious for construction of business centers in once outer neighbourhoods and large scale housing developments. With worsening economic conditions, decreased job opportunities in construction previously absorbing majority of migrant male labour force, new migrants coming after 1990s do not have this opportunity. In addition to gecekondu population who are still trying to cope with poverty with previously set community networks, today, there is the urban underclass, excluded from both official safety nets and solidarity ties of the early migrants (Erman 2003, 51). Moreover, early migrants' decision to come to the city was more of a well-calculated decision based on availability of ties in the city through relatives or through the people they know previously from their hometowns (named as *hemşehri* in Turkey). However, migrants of 1990s are different. Early migrants have come not only from Eastern and Southeastern Anatolia but also from Central Anatolia for economic reasons. However, tide of 1990s is overwhelmingly from eastern and southeastern regions particularly –in addition to economic factors- because of social unrest due to the armed conflict between government forces and PKK, no matter voluntarily or forced. Therefore, it might be argued that this is more of a sudden and forced migration without having adequate ties in the city that can help to find a job and a house.

Coming to the demographic factors, in these conditions, it is possible to claim that family is much more an unstable social unit. The new poor in Turkey coming from eastern and southeastern regions in Turkey still have high fertility rates, decreasing the chance of each child to have proper growing conditions. Although divorce rates are still comparably low in Turkey in line with the social structure, still lone-parenting has a dimension of poverty and especially poverty for children. Share of all children in lone-parent families is 0.7%, however poverty rate among these children is as high as 29.2% compared with children in poverty in other families as 19.6% (UNICEF Innocenti Research Center 2000, 10). Divorced or not, with harsh economic conditions and lacking social security network, family is no longer a supportive mechanism protecting people from poverty.

Considering child poverty in Turkey contemporarily, particularly the latest migrant group might be emphasized. On the one hand, not only latecomers but also certain groups already residing in the cities suffered from economic difficulties which also lowered living standards of children in these families. Previously available health and education provisions by the state, now withdrawn in majority, increased relative cost of raising children thus decreasing psychological

45

value of children for families and increasing economic value: cost of children not working now means increased cost and loss of otherwise increased family income. In some families, male household heads are usually unemployed or underemployed,[3] women have begun to work frequently in the informal sector (Gündüz Hoşgör 2004, 15). In Turkey, patriarchal values might disapprove women working outside. Persistence of patriarchy in these migrant groups combined with increased relative cost of children not working, it might be argued that the increase in child labour in Turkey might also be relevant to disapproval of women working out of the house.

The increase in child poverty not only increased child labour, but also added new job types children are engaged in, one of which is street work. Referring particularly to the working street children, these newly emerged forms of child labour are no more a skill-investing or contributing to the family budget. They are usually the sole breadwinners in their families as either their fathers cannot find a job or working in the informal sector with low income (Kahveci et al. 1996, 43; Akşit et al. 2001, xi) or no longer want to find a job while their children are already earning income (Gündüz Hoşgör 2004, 15). Child labour is frequently no longer an after-school activity; most of new child workers had attended school, but many of them eventually drop out after they begin to work (Atauz 1989a, 9; Akşit et al. 2001, 39). As the post-industrial economy does not have as many as past opportunities for children, new forms of child labour emerged, working street children being the most prevalent among them. Particularly, the children of the latest migrant families in the cities with patriarchal norms and socially excluded condition have the highest risk for street work.

3.4 Working Street Children in Turkey

Emergence of working street children in Turkey as a form of child labour might be traced back to 1980s. According to research, majority of the families of today's working street children migrated after 1985 (Karatay et al. 2000a, 441; Kahveci et al. 1996, 40; Akşit et al. 2001, 64), therefore, 1980s might be referred as the beginning date of emergence of working street children in Turkey. The rise of poverty and child poverty increased economic value of the children, the

---

[3] Working men in the poor families have low income. Many of them cannot find a job for long periods and the jobs are not permanent. In a study conducted on working street children in İzmir, one child states that "My father is unemployed. He is a construction worker looking for a job." Another one says "He cannot get a job permanently. He sometimes goes out of İzmir to work in construction there". (Kahveci et al. 1996, 43).

cost calculated relatively by adding increased costs of raising children and their not working (Karatay et al. 2000b, 473). Other forms of child labour emerged in this process, such as child employment in furniture sectors or seasonal agricultural work. However, the newly emerged forms are the transformed versions of traditional types of employment available to children. On the contrary, working street children are totally new, depending on the dynamics relevant to transformation of poverty into a new kind of poverty.[4]

In Turkey, majority of working street children belong to the first category of children referred in the previous chapter, who work on the street but somehow spend the night at home and are in touch with their families (Akşit et al. 2001, x). Their ongoing contact with their families are supposed to provide, at least to an extent, family protection (Atauz 1989a, 35).

The second group composed of those who also spend the night on the street, is mostly involved in garbage collection and separation; many of them tend to engage in illicit activities such as drug abuse, street gangs and prostitution. As passing to scavenging often brings drug addiction through glue/thinner sniffing (Akşit et al. 2001, 43); it might be claimed that breaking up the ties with the family and entering to this second group might be in relation to the tasks performed.

In terms of terminology, in Turkey, a further distinction is made particularly by the General Directorate of Social Services and Child Protection (*Sosyal Hizmetler ve Çocuk Esirgeme Kurumu – SHÇEK*), as "registered working street children" and "unregistered working street children" based on the criteria whether they are registered to available "Child and Youth Centers" (SHÇEK et al. 2005, 6). Currently, there are 44 child and youth centers in Adana, Adıyaman, Ankara, Antalya, Aydın, Batman, Bursa, Çorum, Denizli, Diyarbakır, Düzce, Edirne, Elazığ, Eskişehir, İstanbul, İzmir, Kayseri, Kocaeli, Konya, Kütahya, Malatya, Manisa, Mersin, Sakarya, Samsun, Siirt, Şanlıurfa and Yalova and 5 monitoring houses in Ankara, Antalya, Diyarbakır and İstanbul (SHÇEK 2006).

Finally, it is to be noted that "working street children" is considered as a "worst form of child labour" that is to be eliminated immediately. In the Time-Bound Programme of Turkey under IPEC, together with seasonal commercial agriculture, informal sector work and domestic labour, street work is considered a worst form as the children on the streets (Dayıoğlu and

---

[4] Both Atauz (1996) and Karatay et al. (2000b) state that working street children exist in Turkey since 1950s. However, it is also explained by these authors that today's working street children are different than the children on the street in 1950s in terms of both quantity and characteristics (Atauz 1996, 467).

Gündüz Hoşgör 2004, 14). Tendency of the children working on the street to be involved in illicit activities such as prostitution, drug dealing and child trafficking, combined with physical risks due to unhealthy working conditions, is resulted in considering working street children among the worst forms of child labour by ILO/IPEC (Ministry of Labour and Social Security 2005, 38).

3.4.1 Factors behind Emergence of Working Street Children in Turkey

Factors behind working street children are multidimensional. The model below is developed by Gündüz Hoşgör for analyzing the causes and consequences of children working on the street (Gündüz Hoşgör 2004, 17). The model is constructed by taking into account the multiple factors behind emergence of working street children on the international, national, local and family levels. Reference of the model of Gündüz Hoşgör to the international dimension might also be thought as establishing a link between working street children and New Poverty in that it points to the role of global forces which was previously discussed in the transformation of poverty.

According to Gündüz Hoşgör's model in Figure 3.1, (Gündüz Hoşgör 2004, 17), the factors influential on the rise and transformation of child poverty in Turkey are also accountable in the case of working street children, poverty being the main underlying factor.

48

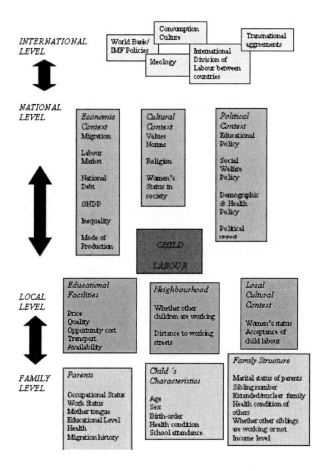

Figure 3.1 Model of Gündüz Hoşgör for Identifying Causes and Consequences of Children Working on the Streets

Emergence of working street children might be understood as a combination of factors deriving from cost of global dynamics as increased unemployment and harsh living conditions with reduced protection and services on behalf of the state. On the international level, Gündüz Hoşgör's model refers to the policies of World Bank and IMF which are previously referred under structural adjustment programmes. The underlying ideology of these policies is neo-liberal paradigm establishing the theoretical framework for implication of these policies particularly involving reductions in social expenditures. One of the most intriguing aspects of the model is that it refers to the role of consumption culture and international division of labour between countries and transnational agreements (Gündüz Hoşgör 2004, 17). On the national level, model refers to economic, cultural and political aspects. On the local level, educational facilities, neighbourhood and local cultural context is taken into account. Finally on the family level, parents, child and family structure is referred on the basis of their implications on working street children.

Here, it is to be noted that it is not the absolute poverty, but rather unequal distribution of income and relative increase in poverty which determines poverty resulting in such an occupation like street work (SHÇEK et al. 2005, 1). Interpreting on the basis of Gündüz Hoşgör's model, the rise of consumption culture in deepened poverty conditions increased expectations of families economically and this played an important role in their decision of sending their children to the street (Udry 2003, 13). Influence of neo-liberal ideology on the policies resulted in cutbacks in available social policies, further complicating the problem. International division of labour switched labour-intensive jobs to developing countries including Turkey, which in turn means available ill-paid jobs without any security, creating working poor in need of extra income from child labour.

On the national level, scrutinized in previous sections, the process created pockets of poverty in terms of economic context. Culturally speaking, there is public approval for children working in order to contribute to the family budget and learn a skill that might be useful in future employment (Akşit et al. 2001, 68-69). Historically speaking, working of the children is also seen as a method of socialization and transfer of knowledge and skills (Atauz 1989a, 7). It might be said that this approval is more of an unawareness that these children are not contributing, but rather earning as breadwinners (Kahveci et al. 1996, 38), no skill investment is made, rather children are deprived of physical and mental development as well as education. Traditional approval of children working in order to contribute to household budget in poor families might be claimed to continue in the case of children working on the street. Statements by the children such as "my father asked me to earn some money", "my mother asked me to earn some bread money. My father made the shoe-shine box" imply that families not only approve their children's

50

working, but also encourage (Kahveci et al. 1996, 56). This kind of approval on behalf of families might be assessed as the families are not aware of the fact that working on the street is beyond earning for the family well-being; but it involves quite dangerous dimensions such as sexual abuse (Karatay et al. 2000a, Küntay and Erginsoy 2005); illness, violence and accidence (Karatay et al. 2000b, 539). In addition to unawareness and traditional attitude towards children working in order to contribute to the household budget, families' approving their children's working on the street might also be discussed within the context of decreasing psychological value of the child in the poor families. Karatay et al. (2000b) states that the tasks that are performed by the children on the street, for instance shoe-polishing or selling items such as napkins and chewing gums, could also be performed by the parents; however, parents prefer to send their children to work on the street. The authors explain this situation with the loss of the "value" the poor families have for their children (Karatay et al. 2000b, 473).

Persistence of patriarchy preventing women from working outside might be considered as a contributing factor for working street children. It should also be noted that labour market also provides fewer jobs for women even if they might work. Fewer opportunities in the labour market even caused a decrease in female participation in the labour force (World Bank 2006b, 4-7). On the political level, increased cost of education and health, and shrinking social welfare provisions is another dimension (Gündüz Hoşgör 2004, 17).

All these factors have decisive roles, however, it is to be underlined that migration, mostly due to the social unrest in the eastern and southeastern regions, is the most persisting factor for why these children could not find better jobs through community networks and remained without protection having to work on the street. Migration to the cities might be argued to affect children negatively also because they are deprived of the support of the tradition family structure in the countryside (Atauz 1990a, 5). Earthquakes of 1999 is an accompanying factor from this perspective pushing many children working in industrial settings or informal sector to the streets as the other opportunities lost with the earthquake, their families usually migrating to other cities without any work opportunities or community networks (Gündüz Hoşgör 2004, 14).

On the local and family level, high price and low quality of education, high fertility rates resulting in crowded families are among the other factors (Gündüz Hoşgör 2004, 17). Mother's language skills, whether she speaks Turkish or not, emerges as an important dimension as an

51

important portion of the mothers of working street children is reported that they cannot speak Turkish (Akşit et al. 2001, 62-63).

3.4.2 Scope of Working Street Children in Turkey

Working street children are part of the transformed child labour in Turkey. Therefore, an introductory note on the situation of child labour is significant for the scope of working street children.

The size and dimensions of child labour in Turkey are identified by surveys made by the State Institute of Statistics in 1994 and 1999. These surveys depend on general methodology and definitions of IPEC. According to the results of 1999 Child Labour Survey; 1 million 635 thousand children between 6-17 years group out of 16 million 88 thousand children population, are at work in economic activity (10.2%); and 4 million 785 thousand children are employed in domestic chores (29.7%) (State Institute of Statistics 1999, 45). This amounts to almost 40% of total child population when added. In general, 61.7% of total number at work is boys and 38.3% is girls at work in economic activity. 29.9% of boys and 73.1% of girls work at domestic chores (Ministry of Labour and Social Security 2005, 15) (Figure 3.2).

girls, 38.3%

boys, 61.7%

Figure 3.2 Children at Economic Activity in Turkey (Gender)

Figure 3.3 refers to distribution of working children. Of these children at economic activity, 57.6% is employed in agriculture, 21.8% is in industrial sector, 10.2% is in business and 10.4% is in the service sector (Ministry of Labour and Social Security 2005, 19). Figure 3.4 indicates that between 1994 and 1999, number of children working in agriculture decreased and

52

the number of children in industry, business and service sector increased (State Institute of Statistics 1999, 23). This change is explained by the decreasing employment of children in agriculture in the survey.

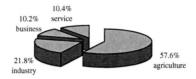

Figure 3.3 Sectoral Distribution of Children at Economic Activity

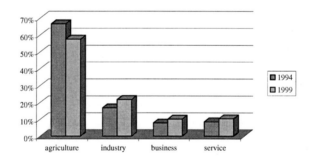

Figure 3.4 Sectoral Distribution of Children at Economic Activity
(1994-1999)

88.3% of total child workforce work for less than 24 hours per week. 1.6% of them experience occupation related accidents and diseases (Ministry of Labour and Social Security 2005, 20). 62.6% of children at economic activity earns less than minimum wage, but contribute to family budget by a 17% increase. The survey identifies contribution to household income, to paying of debts, to helping household's own economic activity, to compensate personal needs, to acquire an occupational skill, and family enforcement as major reasons of child work in Turkey (Ministry of Labour and Social Security 2005, 24). More than half of these children drop school.

The picture gets clearer when compared with the 1994 results. The 1994 survey excludes 14-17 age group. From 1994 to 1999, the total number of children between 6-14 at work in economic activity dropped from 8.5% to 4.2% because of the economic crises and decline in agricultural employment due to the general trend. On the other hand, the percentages for work at domestic chores increased from 25.8% to 28.9% (Ministry of Labour and Social Security 2005, 27). School attendance rate for 6-14 age group increased from 86.9% to 88.1% because the compulsory education boundary has risen to 8 years. School attendance specifically increased in the rural places because the compulsory education obligation is more easily controlled. Extention of compulsory education to 8 years increased general school attendance rates in general. However, in the urban places this trend is reverse. 92.7% school attendance ratio of boys decreased to 91.9%, and 89% girl school attendance fell to 87% because of the increased economic difficulties in the urban areas.

The available studies on working street children are based on the children which could be reached, therefore it is difficult to have an aggregate number for working street children. Indeed, Karatay et al. (2000b, 457) emphasizes that although number of children on the street is important, it is even more important to know that it is not possible to know the exact number, therefore, the studies should emphasize policies and the need for institutionalized efforts. Some resources claim that nearly 40.000 children are living on the street in Turkey, and one in seven children is faced with the risk of meeting street as workplace (*Cumhuriyet* 10.02.2005).

Gender relations are among significant factors in shaping the nature and functioning of child labour (ILO 2003, v). According to research findings, average age of children on the street is 12, majority of them being boys (Gündüz Hoşgör et al. 2005; Kahveci et al. 1996; Atauz 1996; Karatay 2000a, 2000b). Average age for boys is also higher than the average age for girls (Karatay et al. 2000b, 476). In line with patriarchal social values, girls are not sent to street for work or if sent, pulled back from the street after puberty (Karatay et al. 2000b, 476). In the earliest studies on working street children, it is even stated that "A great majority of the children working on the street and almost all the children of the street are boys, with the exception of beggars and gypsies" (Atauz 1989a, 39). The same study also attaches role to religion in that Islam "inhibits the exposition of girls to the street" (Atauz 1989a, 39). Nevertheless, in my opinion, as all of the three big religions put normative roles for women and girls, influence of the patriarchal family structure might be more effective than religion.

Distribution of children working on the street in terms of gender changes in accordance with the location. While percentage of girls is around 6% in Diyarbakır (Atauz 1997), this percentage is 9% in İstanbul as a whole (Karatay et al. 2000b), and it rises to 37% in Beyoğlu district (Karatay et al. 2000a). Therefore, decreased power of patriarchal values might also increase percentage of girls working on the street. Patriarchal values might keep the number of girls working on the street low, however, girls are the first to be taken from school and they do domestic work, and take care of the younger siblings in the house.

In terms of family background (Gündüz Hoşgör et al. 2005, 38), the children belong to migrant families coming from Eastern and Southeastern Anatolia after 1990. The families are crowded particularly due to high fertility rates in the origin cities (in year 1990, percentage of 0-14 age group in GAP region is 47%) (Southeastern Anatolian Project (GAP) 2000). Average number of children in the family is found as 6 in the study on working street children in İstanbul (Karatay et al. 2000b, 478).

Parents have low education levels in general. Mothers are mostly illiterate (72% in İstanbul) and do not speak Turkish (Karatay et al. 2000a, 442). Fathers are mostly literate, but only one third of the fathers completed elementary school (Karatay et al. 2000b, 514). Almost 90% of mothers in İstanbul state that they are "housewives", others are cleaning other people's houses, working in the textile workshops or selling items such as flowers and napkins. Fathers are frequently unemployed (40%) or working in the low income jobs such as construction and peddling (Karatay et al. 2000b, 513). Families mostly live in the squatter settings, however, some of the most recent migrants live together with other families in the same house, each family renting one room, or they live in tents or huts (baraka) (Atauz 1996, 469).

The children are mostly occupied in shoe-polishing, selling petty commodities such as napkins, water, chocolate, stuffed mussel, chewing gums, toys, candies and religious books; cleaning cars/windscreens of the cars and weighing (tartıcılık) (Karatay et al. 2000b, 494). Some of these commodities are home made (prepared at home such as boiled corn, pastry and pie)

(Atauz 1996, 470). Some children are also engaged in begging (Zeytinoğlu 1989, 16). They might also be involved in more dangerous activities such as scavenging (garbage collection).

The tasks the children are engaged with might change with the season, age and city. There is also a tendency among the children to begin with selling petty commodities or shoe-polishing; but later switch to scavenging as this task is more profitable (Akşit et al. 2001, 40-43). The commodities they sell arc provided from various sources including wholesalers. Depending on the city they are living, the children usually do the same task for earning; however, as it is mentioned before, some of them switch to scavenging in time (Akşit et al. 2001, 42-43).

Most common task of earning might change according to location. For instance, while selling napkins is the most common task in Beyoğlu district, shoe-polishing is the most common task in İstanbul in general (Karatay et al. 2000b, 495). Children usually work in the crowded places such as city centers, bazaars, landing stages and subway stations (Atauz 1996, 471). Mainly the jobs preferred by children are those which do not require much capital, skill and experience. In addition to these characteristics, the jobs which do not need continuity of work are also preferred as they might handle the job with the school or they can switch to a more profitable task (Atauz 1990b, 24).

Another way of earning on the street, particularly in the case of children living on the street is prostitution. Family structure is identified as the most important factor for ending up at street by the girls. It is claimed that girls are attached lower value in the family, and main roles are identified as performing household tasks and getting married usually in exchange for a certain amount of money (*başlık parası*) without their consent. These factors combined with the extreme value attached to virginity lead to emergence of street as a way to be away from the family environment (Küntay and Erginsoy 2005, 16). Alcohol addiction and in a few cases drug addiction exists in the family, especially among fathers. Most of the families are migrants and average number of children is over 4 (Küntay and Erginsoy 2005, 96-97). In some neighbourhoods, migration pattern is different in that, for instance in Aksaray and Laleli districts, some women and girls coming from former Eastern Bloc countries are engaged in prostitution (Küntay and Erginsoy 2005, 129).

Education level of parents is usually low. Although all fathers of the girls in the study are literate, 60% of them are either completed or dropped out elementary school. None of the fathers went to university. 10% of mothers are illiterate and 50% of them graduated or dropped out of

56

elementary school (Küntay and Erginsoy 2005, 99-100). 10% of the fathers are deceased, 13% is unemployed, around 10% of fathers are working in the public sector; however, the rest is working as self-employed, workers and street peddlers. 40% of the mothers are housewives, the rest is working as workers, cleaning other people's houses, and working on the farm. 6% of the mothers are working in the public sector and another 6% is working in the entertainment places as "consommatrices" (Küntay and Erginsoy 2005, 101). Only a quarter of the parents are married, both parents being the biological parents of the girls. Death, divorce and marriage to other people are common among the parents. This finding might indicate strong influence of family disorganization on ending up at the street, maybe not for working street in general, but for children living on the street.

The study notes that patriarchal family structure has important repercussion in terms of underage prostitution in Turkey, however, this situation is not specific to Turkey and violence takes place worldwide. The problem with underage girls (girls below 18 years old) in prostitution is explained in Turkey and İstanbul as the priority provided by the capitalist economic system to the hierarchical heterosexism[5] which is historically developed and shaped in the other patriarchal systems and family (Küntay and Erginsoy 2005, 53). The same understanding labels the girls in prostitution as bad although the customer is still respected and seen as a natural activity on behalf of the men. A participant of the study tells the violence one of the customers exerted on her and when she went to the police station, the police officers blamed her and said "who are you so that you can complain about a respectful businessman?" (Küntay and Erginsoy 2005, 57).

The girls might run away to the street and eventually be abused by the people for sexual purposes. It is also the case that girls do not directly engage in prostitution. Some girls working in the informal sector - particularly the girls working in textile sector and hairdressers are identified in the study – might come across people who are themselves engaged in prostitution or making profit out of other women (Küntay and Erginsoy 2005, 25). This does not mean that similar occupations in the informal sector necessarily end up in prostitution, however, the study assesses that these jobs are common as the previous working places of the girls. Majority of the girls interviewed stated that they were working in the textile sector, some of them in hairdressers, in

---

[5] Hierarchical heterosexism is defined as predominance, prioritization and approval of heterosexual intercourse in the society in line with hierarchical power structures in the capitalist system which emphasize superiority of men, rich, stronger (physically and in terms of financial power) and older and allow them to exploit others (Küntay and Erginsoy 2005, 52-43).

shops as sales assistants and others were working in the factories (Küntay and Erginsoy 2005, 91). For the girls having problem in their families and trying to earn money, with convincing statements of mediatory people or other sex workers; prostitution emerges as a job which requires "no skills" other than being women (Küntay and Erginsoy 2005, 124).

Another factor that lead to increase of prostitution is related to urbanization patterns in Turkey. Rapid urbanization and mass migration expanded the big cities in Turkey (especially İstanbul, İzmir and Ankara), and settling areas expanded the cities to the peripheral areas. These areas are also identified among the main places for the girls engaged in prostitution and their "customers" although the places where the girls and customers (and the middlemen) meet for the first time might be in the city centers and entertainment places (Küntay and Erginsoy 2005, 32). Increased relocation of prostitution to the periphery is explained by the more strict legal regulations. In the central places, prostitution is also moved towards the secondary districts. For instance, gambling and prostitution is more intensive in Cihangir and Tarlabaşı which are the secondary circles around Beyoğlu. The negotiations and meeting with the customers might occur in some houses, hotels, restaurants, poolrooms and hairdressers; or in forests, parking lots, exits of oil stations, bridges and highways (Küntay and Erginsoy 2005, 129).

The people involved in the exchange are the girls in prostitution, dealers, other mediatory people who convince, deceive or force the girls for prostitution and the customers. Sexual intercourse is transformed into a service and commodity in transaction (Küntay and Erginsoy 2005, 60).

The girls interviewed in this study are between 14-18 years old, however, as some of the girls state that they are engaged in prostitution for 5 years, age of beginning prostitution is quite lower. Most of the girls were 11-12 years old when they first involved in prostitution (Küntay and Erginsoy 2005, 87). Majority of the girls are literate, however, their education level is low in general. Only one third of the girls completed elementary school, and one third of them dropped out secondary school. Reasons for drop out are economic difficulties, reluctance on behalf of the girls (for mistreatment at school and attractiveness of the city), parents' not sending to school and poor performance (Küntay and Erginsoy 2005, 88-89).

How a child spend day-time might change according to location, age or the task for earning. According to a study, majority of the children wake up between 7 and 11 a.m. Only the children who work at night are reported to wake up in the afternoon (Karatay et al. 2000b, 523).

Majority of the children do not have breakfast at home and eat savory rolls or pastry on the street. Most of them also have lunch and dinner again with rolls and pastry. More than half of them return home as late as 9 p.m.; however, one fifth of the children stay on the street until midnight. Children usually have one day for resting, usually Sundays as there are fewer people on the street whom the children can sell items. In the leisure time, most of the children state that they play games, and a small portion does homework (Karatay et al. 2000b, 527). Although majority of them say that they like working, one third of the children do not want to work as they are getting tired and they would like to go back to school (Karatay et al. 2000b, 529).

Many of them smoke; and mainly those living on the street, particularly garbage collectors use drugs like sniffing glue. According to a study, 17% of these children dropped out of school after completing elementary school, 25% did not complete elementary school and 13% have never attended school (Gündüz Hoşgör 2004, 6). Education level of working street children might also change in accordance with the location. For instance, in İzmir, 22% of the children graduated from elementary school, but did not continue his/her education, 18% dropped out of elementary school and 19% of the children have never attended school (Kahveci et al. 1996, 39). The work on the street causes harms to their physical health due to sanitation difficulties and infection. It is also the case that many children, particularly the children of the most recent migrants living in the rented rooms, tents or huts, suffer from illness as these houses might not have adequate water facilities or sewage system (Atauz 1996, 469). Most of the children do not eat enough and properly. Especially the children working in scavenging are malnourished, and they eat the food they get from the garbage. 72% of the children complain of fatigue (Akşit et al. 2001, 53). Mental harms are observed and although most of them are not deviant initially, after working on the street, socialization problems develop due to abuses and violence.

In the following part, the response measures taken in Turkey against the problem of working street children will be discussed as an attempt for a better understanding of the context of working street children in Turkey.

3.4.3 Response to the Problem of Working Street Children in Turkey

In Turkey, the first legislation about the working children (law dated 10.09.1921 and No. 151) sets minimum age for working in galleries was determined as 18 in Zonguldak coal mining area. After this local regulation, two general laws about working children were initiated as the Law of Obligations (1926) and the Law on General Health Protection (1930).

The Law of Obligations includes some rules on the protection of the rights of apprentices. One of these is "Apprenticeship Contract Agreement" that still prevails despite having a limited scope of implementation (Article 318). Article 330 of the same document ensures apprenticeship training and excluding nights and weekends from the working hours of apprentices.

The Law on General Health Protection is about public health. It has a special section titled "Protection of Children" specifying the minimum age of 12 for working children even rejecting participation of a child under this age into "imece", the local name for the collaboration of village people. This regulation also forbids, children between the ages 12 and 16 to work more than eight hours in a day.

The Labour Law was accepted in 1936, and the "Regulations on Hard and Dangerous Work" based on this in 1940. According to this regulation, children under the age of 18 cannot be employed in work constituting a threat to their safety and damaging their health, as listed in the attachment. Several amendments have been made in the Labour Law at various dates. With the amendment dated 22.5.2003, based on the law no. 4857, a provision stating that children under age of 15 are not allowed to work, was added to the Labour Law (in special cases 13 years old). School hours for children attending to school are included in their working hours by Article 67. In Article 69, it is stated that children under 18 are not allowed to work underground, under water, and during nights. Medical examination before starting the job and every 6 months thereafter (Article 80) are obligatory for those under 18.

The Apprenticeship and Occupational Training Law (5.6.1986, no. 3308), accepts children who have completed obligatory elementary education (in 1998, this was extended to 8 years from 5) as apprentices and apprenticeship students. The children are not allowed to begin apprenticeship before 15 or after 19 years old (special cases allow 13-14 years old) (www.fisek.org.tr).

In 1990s, as a response to the changing nature of traditional child labour, a "National Congress on the Child Policy in 1990s" is gathered in Ankara (May 26-27, 1989). The proceedings of this congress laid the basics of the national programme against child labour. In 1992, Turkey is included in the IPEC programmes together with Brazil, India, Indonesia, Kenya and Thailand. The programme has a tripartite approach: government, employers and workers.

60

Civil society groups such as labour unions, trade unions, professional institutions, employer unions and universities are also included.

On the government level, Ministry of Labour and Social Security is the nationwide coordinating unit, Ministry of Education, State Institute of Statistics, General Directorate of Police-Department of Security, Division of Child Protection; on the employer level, TESK (Confederation of Turkish Tradesman and Handicrafts), TİSK (Turkish Confederation of Employer Associations), Turkish Research Institute of Small and Medium Sized Enterprises and Crafts; on the workers level, the TÜRK-İŞ (Confederation of Turkish Trade Unions), DİSK (Confederation of Progressive Trade Unions), HAK-İŞ (Confederation of Real Trade Unions) are included. Fişek Institute, Human Resources Development Foundation, universities and media are also involved on behalf of the civil society. IPEC plays coordinating role among these parties.

Among governmental institutions General Directorate of Social Services and Child Protection (SHÇEK) is very important, since it is one of the basic institutions providing support to the children. According to SHÇEK statistics (SHÇEK 2005a), in Turkey there are 9096 children (boys and girls between 0-12 years and girls over 12 yet still deprived of family protection are defined as "children in need of protection" by SHÇEK and they stay in these institutions) in 86 child nurseries (*çocuk yuvası*) working under SHÇEK. 10509 children in need of protection aged between 13-18 years are currently staying in adolescent houses (*yetiştirme yurdu*). Around 500 children aged between 0-6 whose families cannot take care of them either because of death, being in prison or in economic difficulties are staying in day nurseries (*kreş ve gündüz bakımevi*). And by January 2005, there are 41982 children who had benefited from child and youth centers (*çocuk ve gençlik merkezleri*). Of these children, around 1000 children working on the street have been withdrawn from the street by cash transfers, around 12000 children returned to school (SHÇEK 2005b). Children who have lost at least one of his/her parents, children whose parents are unknown, children abandoned by their families, and/or children who are leaded to prostitution, begging, drugs or alcohol by their families are defined as "children in need of protection" by SHÇEK in the same resources.

SHÇEK adolescent centers (*yetiştirme yurdu*) also give services to the children exposed to sexual exploitation for commercial reasons. Among the girls interviewed in the study on teenage sex workers in İstanbul, some of the girls stated that they stayed at these centers previously, however, ran away to the street together with some other girls (Küntay and Erginsoy 2005, 180).

Some of these girls would like to return to the centers, while part of the others does not want to be back.

By 2001, children abandoned by their families and taken under supervision of SHÇEK are claimed to be over 3500 (Kurt 2001). Economic difficulties, family disorganization and negligence are identified as the causes behind child abandonment. According to a report by Çocuk Vakfı in 2000, it is estimated that there are around 700.000 desolate children (Çocuk Vakfı 2000).

Government institutions conduct surveys, delineates the national policy, perform necessary legal amendments, and in general coordinate and formulate policies under the supervision of Advisory Board and National Steering Committee. Employer and worker organizations are dealing with awareness raising targeted training programmes and civil society groups, they are implementing small scale, location and sector specific projects.

Turkey had signed membership agreement of ILO in 1932, since that time it has signed 1989 UN Convention on the Rights of Children, 1989 Council of Europe Convention, and several ILO agreements including Conventions No. 138 and 182. 101 action programmes were implemented in these 12 years. Around 50.000 children were accessed in this framework. 60% of these children have been withdrawn from work and enrolled in school. The working conditions, health, nutrition and vocational training of the rest were improved (www.ilo.org/public/english/region/eurpro/ankara/programme/ipec/about.htm).

Major strategies of the national programme was initialized in 1992-93 biennium, and enhanced afterwards. These strategies have involved developing a greater understanding of the problem, increasing awareness of the policy makers, undertaking small scale easily replaceable direct action programmes focused on priority areas of concern, expanding these activities and mainstreaming child labour issues into government policies, programmes and budgets.

Turkey is now in the National Time Bound Programme in the process of creating policies and programmes to eliminate the worst forms of child labour in 10 years. The main objectives for 2004-2005 period are contributing to the effective implementation of ILO Convention No.182 and TBP and mobilizing mechanisms and resources to combat child labour. For these main reasons, technical cooperation programs are to be developed and workshops, seminars and symposiums are to be held (http://www.ilo.org/public/english/region/eurpro/ankara/programme/ipec/strategy.htm). In

Turkey, it is directed by the Ministry of Labour and Social Security. IPEC backs the government with additional financial resources and technical assistance. Fundamental elements of the programme is government commitment, linking child labour action to poverty alleviation and accessible quality basic education, rapid response measures for prevention, withdrawal and rehabilitation of victims of the worst forms of child labour, income support to their families, education of children, social mobilization and public opinion activities. *Red Card to Child Labour Campaign* is the benchmark of this programme.

In Turkey, many research activities have been undertaken to identify the worst forms of child labour. Participation of Turkey in the IPEC framework might be claimed to increase research and activities on working street children, however, response to the problem of child labour and working street children dates back to the period before 1990s.[6] In 1989, the first field research on child labour and street children is conducted by Sevil Atauz in Turkey. Atauz also studied on working street children in Şanlıurfa (1990a), in Ankara (1990a, 1990b), in İzmir (1995) and in Diyarbakır (1997). Betül Altıntaş also conducted a research on working street children in Ankara (2003). Karatay et al. (2000a, 2000b) focus on the children in İstanbul; and Kahveci et al. (1996) explore shoe-shine boys in İzmir. Küntay and Erginsoy (2005) conducted a research on girls engaged in prostitution.[7]

Some of these studies are either conducted or published by the financial and technical support of NGOs in Turkey point to the involvement of NGOs in the response to working street children. For instance, studies of Karatay et al. (2000a, 2000b) are published by İstanbul Çocukları Vakfı. The research of Atauz (1995) on the working street children in İzmir is conducted with the financial support of Sokak Çocukları Koruma Derneği (Atauz 1996, 469). Çocuk Vakfı is also engaged in reports on the rights of children and refer to working street children among the children who are deprived of the rights the children enjoy worldwide (2000).

---

[6] The studies referred at this section are the studies which were conducted within the IPEC framework. However, researchers of these reports, for instance Sevil Atauz and Abdullah Karatay, also base their methodology on categorizations and approach of international organizations dealing with children such as UNICEF. Here, only the studies focusing on working street children are mentioned as the discussion is on working street children. However, the studies in Turkey are not limited to these research. There are studies on child labour in Turkey in general and children working in other sectors. Atauz (1989b), Balamir (1982), Çocuk Vakfı (2000), Ertürk (1994), Ertürk and Dayıoğlu (2004), Köksal and Lordoğlu (1993) and Konanç (1996) are among these studies.

[7] The sub-section on the literature on working street children in Turkey refers to the basic points in these studies

Activities of the NGOs are not limited to supporting research on the other hand. They formulate and apply projects targeting specific groups in the areas they are established.

Fişek Institute publishes a journal (*Working Environment – Çalışma Ortamı*), has a media project in order to increase public awareness on children through press and media (*Voice of Children*), provides health service for the children working in industrial sector (Girl workers under 15 years old in Denizli between 1996 and 1999) (child workers in Pendik industrial site), established a social center for working young girls in Ankara (*Genç Kız Evi*). The institute has also projects on security at the working environment and organizes meetings and workshops for awareness raising and training of the child workers and employers (www.fisek.org.tr).

İstanbul Çocukları Vakfı has established a Child House (*Rotary Çocuk Evi*), publishes books on children and is involved in organizing relevant meetings. Umut Çocukları Derneği is established and run by Yusuf Kulca who, himself, was a street children. This association has established First Step Stations in İstanbul aiming re-integration of the children on the street to the society (www.umutcocuklari.org.tr). It has also a school project targeting street children; a laundry, a house for street children to spend the night, projects aiming to establish dialogue with street children. Donor mapping project mentions 33 associations and 44 foundations on children in Turkey (UNICEF 2002).

In terms of children living on the street, another institution should be mentioned. After the research conducted by Küntay and Erginsoy, with the initiatives of the researchers and support from SHÇEK and other governmental institutions, a shelter house for the girls engaged in prostitution is established in İstanbul (Küntay and Erginsoy 2005, 181). Although protection of the girls exposed to sexual exploitation for commercial purposes is under the responsibility of SHÇEK, and the girls reached by social servants are placed in adolescent houses (*yetiştirme yurdu*); co-existence of these girls might have influence other children in the institution negatively (Küntay and Erginsoy 2005, 180). The institution provides services as a First Step Shelter House to the girls engaged in prostitution or under the risk of prostitution due to sexual abuse.

Labour and employer unions are also part of the efforts in Turkey. Activities of these in Turkey (Türk-İş, DİSK, HAK-İŞ, Tisk, Tesk) target both child labour in general in Turkey and working street children. They are involved in awareness-raising among children, families and employers through meetings, conferences and publications, contributing to the efforts of placing

64

working children into the boarding schools (*Yatılı İlköğretim Bölge Okulları - Regional Basic Education Boarding Schools, Pansiyonlu İlköğretim Bölge Okulları – Basic Education Schools with Pensions*) and directing the working children to the school, improving of working conditions, providing apprenticeship training, conducting studies and project for income generation for the families (UNICEF 2002, 18-23).

In the urban places, the children working on the streets (Akşit et al. 2001), cotton gathering in rural areas are chosen as the worst forms. At the governmental level, the project aims at providing the government with necessary information to adopt policies in the relevant areas in order to combat child labour, especially the worst forms of child labour through a national time bound policy and programme framework. Specifically speaking, most recent activities are in Diyarbakır (by the Southeastern Anatolia Regional Development Administration and Directorate of Social Services and Child Protection), in Denizli, İstanbul and Çorlu (by DİSK), in Gaziantep, Karaman and Adapazarı (HAK-İŞ), in Yalova (by Province of Yalova and Ministry of Labour and Social Security), and in Gölcük and Adapazarı (by Directorate of Social Services and Child Protection). Local boarding schools projects of TÜRK-İŞ, DİSK, HAK-İŞ and a time bound programme in İzmir targeting elimination of child labour in İzmir by 2003 are the other projects that might be mentioned among the activities.

In general, the research and activities made within the period after participation in the IPEC framework might be assessed as successful in reaching children on the street and returning some of them back to school in Turkey. In my opinion, this success might be related to the commitment of the participating institutions and organizations both in the public and private spheres. IPEC framework might be evaluated as more of a guideline, therefore, the degree of achievements might change from country to country.

The framework aims to incorporate government institutions with the schools, unions and civil society institutions for eradicating the worst forms of child labour and improving working and living conditions of the other children. Although one of the main aims is to mainstream the policies targeting child labour and working street children into the broader social policies, majority of the projects are on the micro level dealing with a certain group of children in a certain area (or city/district). Devising and developing social policies targeting the children on the national level is very important for the ultimate aim of eradicating especially worst forms of child labour. Turkey has ratified relevant international conventions regarding children, period of the

65

compulsory primary education is extended, issues concerning children are integrated into the 8th Five Year Development Plan; however, poverty and child poverty is still high in Turkey. 37% of children under 15 years old (UNICEF 2006) and 25.6% of the population in general is living in food and non-food poverty (State Institute of Statistics 2004). Therefore, integration of projects regarding children to the national social policies and success on the national level might be claimed to be limited. Without social policies targeting family poverty and child poverty directly, the success of the projects and programmes might not be extended all working street children.

An assessment of IPEC programmes in Turkey states that income generation activities for the families of the children are included in the projects, however, this dimension is not yet a main component (Dayıoğlu and Hoşgör 2004, 46). The focus on the families of the children might, therefore, be claimed to be another shortcoming of the programme. Involvement of NGOs are very important to reach the families (Dayıoğlu and Hoşgör 2004, 46), however, in my opinion, unless measures are taken on the governmental level, achievement of the programme might remain on the micro level and may not prevent the increase in the number of working street children as poverty of the households would not be overcome.

Another shortcoming of the IPEC framework in Turkey might be discussed on the education dimension. Efforts in terms of education should be more comprehensive. The factors which cause the children drop out school or have poor performance, such as poverty, cost and quality of education and cultural dynamics (particularly affecting girls) should be explored in detail and ongoing studies in terms of education both on the research and project level should be increased and developed (Dayıoğlu and Hoşgör 2004, 39). IPEC framework is a general international framework and with the research conducted in individual countries, this deficiency is tried to be overcome, but it might still be considered limited as the report on the evaluation of the programme mentions (Dayıoğlu and Hoşgör 2004, 39). Therefore, together with the need of increased corporation between governmental, civil society and international organizations, studies on the context of child labour in Turkey should be increased and programme should be developed on the basis of specific factors that lead to emergence and increase of child labour and working street children in Turkey.

In this chapter, discussion of working street children in Turkey leads to the interaction of international factors with the national and local factors in Turkey (Gündüz Hoşgör 2004, 17). Decline of absolute poverty accompanied by a rise in relative poverty (Dansuk 1997), and persistence of social exclusion and inequality dimension (Buğra and Keyder 2003) points to transformation of poverty in Turkey as New Poverty. The international dimension and New Poverty in Turkey are accompanied by national factors such as social unrest, migration, education and earthquake (Akşit et al. 2001), all of which might be evaluated as emergence of working street children is a multifaceted issue based on the international and national processes.

CHAPTER IV

WORKING STREET CHILDREN IN ROMANIA

4.1 Introduction

In the previous chapters, firstly the theoretical framework of New Poverty and child poverty is provided. Emergence of working street children as a new form of child labour in this context in relation to transformation of poverty and child poverty is discussed. Then, Turkey is introduced. The situation of children and child poverty is discussed in the Turkish context beginning with 1990s since the general claim is proposed as the post-1990 period is a period in which the mentioned transformations have taken place. The factors that led to the transformation of child poverty which resulted in emergence of working street children are examined. The chapter is concluded with the scope and dimensions of the notion of working street children in Turkey together with the responsive measures.

This chapter has a similar outline trying to link the global process to Romania this time. After the short literature review, child poverty in Romania, particularly since 1990s, will be discussed in terms of its scope and the dynamics behind. The factors leading to increase and transformation of child poverty in Romania will, then, be scrutinized through the analysis of global dynamics and country-specific factors. Discussing the transformation of child poverty in Romania, working street children in terms of relevant terminology used in Romania, the factors that are feeding the notion, its scope and the response to the working street children problem in Romania would be analyzed.

The main aim of this chapter is to gain insight into child poverty and working street children in Romania in the context of New Poverty and provide a basis for comparison of Turkey and Romania in the next chapter.

4.2 Review of Literature on Child Labour and Working Street Children in Romania

On the child labour and working street children, the earliest study I came across is *Children at Risk in Romania: Problems Old and New* by Catalin and Elena Zamfir (1996). In this study, Zamfir and Zamfir mainly focus on the children in institutions and child abandonment. This study is important in that abandoned and institutionalized children form an important portion of contemporary working street children in Romania (Mother and Child Care Institute 2005).

A similar study is by UNICEF Innocenti Research Center entitled *Children at Risk in Central and Eastern Europe: Perils and Promises* (1997). This study also does not specifically focus on working street children, however, gives an idea about the transition of the countries to the market economy and the relevant risks posed to children in this period.

The main data on child labour in Romania relies on the study of Ghinararu (2004) and the survey of Romanian National Institute of Statistics and ILO (2003) on children's activity.

Studies which directly target working street children are conducted in Romania after the country's participation in the IPEC framework. These studies employ a similar methodology with the IPEC framework studies in Turkey, particularly with *Turkey: Working Street Children in Three Metropolitan Cities. A Rapid Assessment* (Akşit et al. 2001). Alexandrescu (2002) focuses on working street children in Bucharest, while another study is carried on Bucharest, İaşi and Craiova cities (Save the Children Romania and IPEC 2003). Data used in this chapter on Romania mainly depends on these two studies. A specific study on *Roma* children (also known as *Gypsy* children) which is carried out by Save the Children Romania (2002), moreover, is used in this study on the parts on *Roma* children. Gregorian et al. (2003) analyzes working street children from justice perspective.

4.3 Child Poverty in Romania

According to the latest *Human Development Report 2005* released by UNDP, Romania is among the medium human development countries together with Turkey (UNDP 2005, 379). It is also expressed as a medium income country in the same report, at the rank of 64 (UNDP 2005, 380).

In 1989, 7% of the population was estimated to be under the poverty line. In 1994, the poverty rate estimations was varying between 22% and 39%. By 1999, poverty rate was 41.2%, which means that poverty increased by 60% compared with four years ago; and by 2000, this rate is claimed to be 44% (Zamfir 2002, 7). Given that these estimations depend on the variables of World Bank and the Research Institute of Quality of Life, the percentage might be claimed to be higher. In terms of relative poverty, the resources on Romania do not provide percentages for relative poverty since the countries with high inequality tend to use absolute poverty levels. Instead, severe poverty rates are given ranging from 12.3% in 1995 to 12.2% in 2000 (CASPIS 2002). The same report claims that the rate of poverty in 2000 is around 30% therefore somewhat higher levels might be predicted for the severe poverty rates. Coming back to relative poverty, the statement that 80% of families with more than three children live in poverty might provide an idea (UNDP 1998). It is claimed that inequality increased by 50% in the initial years of transition and the income of top 5% exceeded by 15 times the income of the lowest 5% (Zamfir 2002, 7). The report providing ranks for countries in terms of under-five mortality rate also refers to the share of household income between 1992 and 2002 stating that lowest 40% gets 21%, while highest 20% gets 38% (UNICEF 2004, 108).

In this general picture, there are controversial statistical data on child poverty in Romania. However, all resources accept that child poverty increased more than the general increase in absolute and relative poverty since children are basic beneficiaries of the social welfare provisions which were reduced considerably after 1990s. For Romania, child poverty rates have increased one and a half times more than the general poverty rate (UNICEF 2003, 3). According to the *Situation of World's Children 2005 Report* released by UNICEF, the under-five mortality rank of Romania is 120, its value being 20 in 2003 (UNICEF 2004, 115). 9% of infants are born with low weight (UNICEF 2004, 122). This is a quite promising percentage compared with other developing countries, however, thinking that the most developed countries in terms of child well-being have at most 4-5 % - for example, Norway has 5% and Sweden has 4% - (UNICEF 2004, 122), it might still be considered as a high figure. Literacy rates for Romania are high, which might be explained with the educational policy of the socialist regime accounting to 99% for males and 97% for females (UNICEF 2004, 134), however, there is a slight tendency of decreasing literacy due to the increasing cost of education and changed educational policy. In terms of child marriage measured as percentage of women 20-24 years of age married or in union

71

before 18, no data is available in the report, though particularly among the children working and living on the street, this percentage is increasing. In 1998, 38% of children under 7 years of age and 50% of children between 7-15 years were living in poverty; the poverty rate of those between 16-24 was 45.5% (Zamfir 2002, 13). Moreover, the percentage of Romanian children at risk of poverty is around 21-23% according to 2004 figures (Ghinararu 2004, 56).

In terms of Millenium Development Goals, there are mostly positive indicators, however, the structural problems caused by the transitional process prevents taking concrete steps. The literacy rate of those between 15-24 corresponds to 99.6% by 2000 (Government of Romania and UN Romania 2003, 11), however, this rate might be misleading in the context of the country's current situation as the age group is not born after 1990. The net enrollment ratios at the elementary level rose from 91.2% in 1990 to 96.8% in 1998 which indicates a general positive tendency in primary education. Yet, other resources claim that due to increased education costs and decreased quality, school enrollment rates tend to decline. Under five mortality rates are decreasing after a temporary increase in mid 1990s.

Statistics and research do not provide much reliable quantitative data; however, an analysis of governmental strategy paper enhances the overview of situation of children in Romania. In the Government Strategy, the children who are at risk or in difficulty are defined as "institutionalized children; children protected in families, both in their own families (in order to reduce the risk of abandonment), as well as in their substitute families/alternate family-type services; children who are maltreated, neglected or abused in their own families; children with special needs and children with HIV/AIDS hosted in institutions or in alternate family-type forms of protection; delinquent children, street children, children/young people coming of age during their long-term institutionalization" (Government of Romania 2004, 9).

In Romania, it can be claimed that the post-socialist period indicates existence of New Poverty. This conclusion can be reached on the basis of previous discussions of New Poverty in the theoretical chapter. As mentioned before, in New Poverty, increase of relative poverty despite possible decreases in absolute poverty might happen. This is a two-fold process in that on the one hand, the opportunities, commodities and services have increased for a certain society in general; therefore the absolute poverty levels might have been even decreased. The increase of economic growth in Romania in the transition period, and the statistical data that the percentage of population living below $1 per day indicates such a process (UNICEF Innocenti Research Center 2004, 2).

However, on the other hand, there are people who cannot benefit from these opportunities. This leads us to a second differentiation, as discussed in the previous chapters that increasing inequality and social exclusion creates pockets of poverty[8] that is more of a permanent character for these people (Zamfir 2002, 31). Therefore, Romania might be given as a distinguished example in this respect. In the post-socialist period, the old days of food and services scarcity diminished drastically with the transition to a market type economy. The opening up of economy have meant that many new goods and services have entered to the country both increasing the available stocks and multiplying their available forms and kinds. In general, availability of the commodities and services increased. Nevertheless, the transition process also meant increasing inequality, regional discrepancy and hardening social exclusions which were this way or other way kept under certain levels in the previous system. Actually, the resources on Romania does not directly point to this fact, they usually state that poverty in general increased significantly, corresponding to 44% in 2002. Yet, references to existence of inequality and social exclusion imply drastic increase in relative poverty as these two concepts explain why poverty for some sections of society increases more. In terms of social exclusion, the references are clearer. During the socialist regime, the social differentiations had tried to be diminished in the state's ideal uniform society project. This was particularly true for the *Roma* community. The costs of transition are not distributed equally, and disadvantaged sections of society are driven into deep poverty. These groups of people are children (particularly the abandoned children and working street children), young people, crowded families, people with lower education and skills, single parent families, homeless people and the *Roma* community (Vilnoiu 2000, 10).

4.3.1 Factors Leading to Increase and Transformation in Child Poverty in Romania

The discussions above point to the fact that Romania's transition in 1990s, also transformed the poverty in the country, thus transforming the scope and dimension of child

---

[8] Pockets of poverty refers to incidence of extreme poverty for some groups in the society despite the general increase in living standards and increased economic growth on the national level (Zamfir 2002, 31).

poverty. In this process, in terms of the factors leading to transformation of child poverty, it might be argued that global dynamics and the effects of neo-liberal paradigm are among the relevant factors. These factors are examined below in the context of the structural adjustment programme and reform measures prepared and applied in line with international financial and monetary organizations such as IMF, World Bank and WTO, the change in the economic structure and social policy provisions, divide generated by structural transformation in the economy and technological innovations, and the rise of consumption culture. After the discussion of international dynamics, the country-specific factors in Romania will be discussed on child poverty.

4.3.1.1 Global Dynamics and Neo-Liberal Paradigm

Going deeper into the motives of New Poverty in general makes the picture clearer. Previously, the effect of neo-liberal paradigm, integration into the global market economy and the overall change in the global political economy, transformation of state structures with decreased social expenditures and diminished intervening role in the economy and finally the change in the society and family form were mentioned as the basic tenets of the international process that resulted in New Poverty.

Concerning Romania, integration into the international political economy opened the country to the international transformation of poverty through the global dynamics that changed the structure of the economic system to a full-fledged market economy in which the role of the state is supposed to diminish in economy, the economic structure changed into a flexible, decentralized, deregulated, privatized and overwhelmingly based on service sector.

In addition to the global dynamics, the worldwide rise of neo-liberal paradigm began to influence Romania's administration through the application of structural adjustment prescriptions of IMF (entered the country in 1991) and World Bank (since 1992). In line with neo-liberalism and structural adjustment programmes, in the reform agenda, the economic growth is prioritized at the expense of social policies. The main justification behind the policies was that economic growth would benefit all sections of society inevitably. The role of the state is diminished in order to provide room for the market mechanisms to function. The previously set social policies were withdrawn. Although this understanding is tried to be replaced with a new understanding which foresees a diminished role for state, but tryies to preserve the social policies to an extent at least in consistency with the requirements of European Union; this problem still goes on today.

Several reform measures are taken in order to liberalize trade. Deregulation of the labour market and flexibility of production are applied as they are among the premises of neo-liberal paradigm. The restructuring produced economic growth on the one hand, however, it produced social costs on the other hand. Unemployment increased with the closure of large scale industrial enterprises and privatization; inflation increased, with the decline in life-long secure public sector jobs, the number of working poor increased as the informal sector emerged in this economic environment with part-time or temporary jobs with few job securities or even without security. The supports and subsidies cut for housing, health and education and the reductions in the child allowances; the living standards of the families declined sharply. Today, the GDP level is lower than it was in the beginning of the transition (Zamfir 2002, 20).

The predominance of the neo-liberal paradigm poses another problem. The harsh criticism of the previous regime had been accumulated for decades, therefore, transition to a new political system and neo-liberal paradigm's rising as a new discourse for the accompanying economic structure, have driven the new state authorities and the bureaucracy to a complete denial of the possibly positive aspects of the previous regime (Dutu 2002; Ghinararu 2004, 27). This approach resulted in abandoning the society-friendly dimensions together with other aspects. Free health, state subsidies in public services, subsidies to housing and schooling can be claimed to be renounced in the new system, particularly in the initial periods, I guess, when the atmosphere of the revolution made people think that a completely different system would certainly be better. Such an approach might be added to the reasons behind increasing poverty, social exclusion and inequality in the Romanian society.

The neo-liberal paradigm has been established in Romania in a similar way many of the developing countries experience, clearly speaking, through the structural adjustment programme (SAP) and reform measures inscribed by international monetary and financial organizations such as IMF and World Bank.

Romania's structural adjustment programme is different from those applied to other developing countries outside the ex-Soviet bloc such as Turkey, as the country is a specific case whose economic structure was overwhelmingly out of the Western mode of economy. However, the basic understanding behind the programme so as to provide a framework in which the in-debt country would make the necessary reforms to improve its economic performance. In accordance with the programme, export-oriented production replaced existing production mode, domestic demand has been intervened to decrease, cutbacks in public spending came into force, state is shrank in many sectors; and deregulation, flexibility, disorganization of labour force and low labour cost are targeted as reform measures (World Bank 2005, v).

Acceptance of neo-liberal paradigm as the new discourse in Romania, also went hand in hand with the global dynamics just as it happened in many other countries including Turkey. It is not possible to claim that Romania was out of the reach of globalization before 1990s. It had responded to the new requirements of the global dynamics previously both by changing and flexing the political economic understanding in time and also by integrating into the international organizations (Romania is part of the UN system since 1955 and joined World Bank in 1972). However, the sudden fall of the existing structure and emerging political economic instability opened up the country to the effects of the global dynamics suddenly and leaving not much time to take measures against the negative effects. Today, most of the elements of globalization in the sense that is described in *1999 Human Development Report* exist in Romania (UNDP 1999, 30).

The country has transcended to a market economy and its national market is increasingly globalized responding to the premises of the global dynamics and opening up its sectors to global competition and investment (World Bank 2006a, 2). The new actors of the globalization particular to the late 1980s, such as IMF, World Bank and World Trade Organization have taken their places in the new system of Romania. Finally, the increased use of technology in almost all sectors and use of Internet as the dominant communication and information mechanism today exist in the country (Government of Romania 2005).

The outcome of this reciprocal relation between globalization and neo-liberal paradigm indicates its existence in the change of the economic structure in line with the global economic process. Industrialization emphasized by the social rule is almost abandoned, large scale industrial production is de-emphasized and sectors of post-industrial production involving the rise of service sector also came into being in Romania (Zamfir 2002, 5-12). Informal sector emerged nearly as a form of underground economy both in the form of adoption to the needs of the new flexible production mode and as a result of the country's economic situation and as part of survival strategies which can not provide employment to many people in the formal sector. The new economy of Romania created with efforts to be in accordance with the global tendencies necessitated removal of national barriers in front of the international trade on the one hand, and on the other hand required flexibility and deregulation in the labour market. The previous full-time, registered, life-long and public sector employments are replaced in many places and sectors by temporary, part-time and sub-contraction jobs (Zamfir 2002, 11).

In Romania, technology is also determining the economic structure likewise other countries in the world, technology-based and capital-intensive jobs are increasing at the expense of labour-intensive occupations. Access to technology creates an additional gap, this time in the form of qualifications and skills, and also in terms of access to information (Niculescu and Admitracesei 2000, 202).

The international factors behind the rise of New Poverty in Romania, and in terms of the new form of child poverty necessitate analyzing the changes in the social policies in relation to the global dynamics on economic structure and the leading administrative discourse, namely the neo-liberal paradigm. Transition to a market economy in Romania and neo-liberal thinking that liberalization and privatization would lead national economic prosperity and competitiveness, all of which were believed to benefit the society as a whole in time, influenced also the understanding behind formation and enforcement of social policies. Chapter 7 of the current government programme in Romania on the Social Security Policy reflects this understanding (Government of Romania 2005). With the cutbacks in state supports and free services, the cost of education, health and housing increased suddenly (Zamfir 2002, 14). Moreover, means-tested benefiting replaced the relatively universalist approach of the socialist regime (Zamfir 2002, 13-14). Increasing unemployment, increased inequality, falling real wages and inflation on the one hand and cutbacks in the available social provisions that might have protected many people against the social costs of transition, summarizes the process leading to high adult and child poverty in Romania in terms of the international factors of poverty (Zamfir 2002, 13).

In this new context of international dynamics, requiring higher skills and education, the youth, the women, the workers previously employed in the labour-intensive industrial state enterprises which have become obscelete today, and children that have to work in order to contribute to the family budget or survive on their own, are experiencing difficulty to find a decent job that can sustain their needs to live (Zamfir 2002, 13-18).

The economic collapse in this process, the decline of wages and increase of unemployment, rise of informal sector after the integration to the international system creating working poor masses, cutbacks in the public expenditures of the state can be summarized as the factors behind the rise of poverty in Romania. With the transition to market economy, inequality increased. As the new system requires higher educations and technological qualifications, wage inequality also increased and this situation has led to emergence of winners and losers in the new period (UNICEF Innocenti Research Center 1999, 91). Increased wage inequality and introduction of disproportionate taxation also increased relative poverty.

Finally, in this atmosphere, the previously controlled ethnic and cultural differences transformed into social exclusion, particularly for the Roma minority (Save the Children 2002, 7). Thus, both increased new poverty incidence for the families; and the cutbacks in education, housing, health and child allowances in line with global dynamics and neo-liberal paradigm through the restructuring transformed and increased child poverty.

4.3.1.2 Country-specific Factors in Child Poverty in Romania

Assessing the causes of a problem or the insights of a fact in a given country necessitates looking at both the factors that are shared by other countries and those specific to that country. Studying a country which is part of the former Soviet bloc, in this respect, requires specific attention on the internal dynamics since their experience with the market economy differs from those of the other developing countries which were increasingly integrated to the global political economy in late 1980s. Romania is among the 27 transition countries from the former Soviet system. Thus, the country-specific dynamics behind the transformation and increase of child poverty in Romania might be discussed within this transition process (Zamfir 2002). Transition process and its implications and social costs are shared by other countries in the same group; however, each country had a somewhat different experience of this period. Therefore, from this perspective, the internal aspect of transition in terms of the legacy of the previous policies might be considered to prevail over the international dimension in Romania (UNICEF Innocenti Research Center 1997, 12).

The impact of transition on child poverty is related to the international paradigm in the sense it is explained above. The new system is established in the framework of this general understanding. However, transition of Romania from a state controlled communist system to a Western style democratic system based on market economy also includes internal dynamics.

78

Among these internal dynamics, the economic crisis which is experienced relatively less severe by other transition countries, the old system's legacy in terms of decreased and distorted welfare dynamics in that period, emergence of factors that prepared the grounds for the post-communist period poverty and structural problems; reform deficits, demographical problems and family change might be considered among these internal factors for the rise and transformation of child poverty in Romania.

Almost all transition countries have suffered from economic difficulties in the initial periods. Some countries such as Poland and Czech Republic have overcome these difficulties and achieved a relatively smooth transition to the market economy (World Bank 2002, 5). However, structural problems with the economy, the old industrial technology and almost obsolete industrial settings, combined with impoverishment starting from 1980s; Romania had a problematic period. The economic growth had been interrupted in mid 1990s (World Bank 2002, xv). Besides, the reform schedule had been full of deficits including neglect of social dimensions (Zamfir 2001, 36). The high birth rate as a result of pronatal policies of the communist regime increased the difficulties for the families to cope with the new environment with increased costs (Ghinararu 2004, 44-45). The family disorganization due to the hardships contributed to increase of child poverty (Save the Children Romania and IPEC 2003, 6). These internal dynamics of child poverty will be analyzed in more detail while discussing the factors behind working street children.

These factors not only increased child poverty but also increased the economic value and relative cost of children when they are not working. As a result, child labour increased in Romania. The difficult living conditions and diminished work opportunities in the cities gave way to urban to rural migration. Therefore, the children increasingly became economically active in the rural areas (Ghinararu 2004, 46). The families which do not have the necessary links with the countryside remained in the city, and their children had to work on the street in majority.

4.4 Working Street Children in Romania

Working street children have emerged as a reflection of child poverty in Romania (Gregorian et.al., 2003, 3). The beginning of the problem is dated as 1993 in many resources. Working street children might not be the most common form of child labour in Romania, however, it is the most visible problem about children, reflecting the inner social problems of the country and identified as one of the worst forms of child labour in Romania by IPEC. It is also important in that it is in connection with other worst forms the children are involved such as prostitution and drug dealing. The main concerns about working street children in Romania identified by the National Committee for Child Protection (NCCP) and National Authority for Child Protection and Adoption (NACPA) (Gregorian et.al., 2003, 13) are the lack of reliable data, social exclusion and inequality (for instance for children with disabilities, children living with HIV/AIDS, children in care institutions, children in detention, asylum-seeker and refugee children, foreign children, children between 16 and 18 years, children from poor households, and children belonging to Roma and other minority groups), drug use, alcohol use and smoking, decrease in participation to education and living in the street permanently.

Romania, as an IPEC partner country since 2000, uses a similar terminology in child labour in general and on working street children in particular. In addition to this terminology, the authorities and research institutions in the country makes a four-fold categorization (Save the Children Romania 2005). The first category of children permanently live and work on the street without any connections with their families, children in the second category work on the street but somehow go back to their families every day. In the third category, there are young people who are on the street at least for five years and began as children working and living on the street and have now become group leaders coordinating the children of the streets. Finally, there are the children living on the street with their families (Alexandrescu 2002, 32). Particularly the cutbacks in housing provisions pushed these families to the street.

4.4.1 Factors behind Emergence of Working Street Children in Romania

The factors behind the emergence and increase of the working street children in Romania mainly stem from the transition process. The subsequent rise of poverty (particularly relative poverty), inequality and social exclusion on the one hand, the structural problems tracing back to the communist period, family disorganization, child abandonment, institutionalization, demographical factors, migration, family attitude towards child work and state responsibilities, decreased social protection and increased cost and economic value of children are among the other factors.

Social protections ad policies of the previous regime prevented emergence of working street children in the pre-1990 period (to begin with, it is clear that in the pre-1990 period there were not much children working or living in the streets. This was due to the social protections and policies of the state. However, with the transition period, on the one hand the economic crises, on the other hand withdrawal of available provisions for children, and in general in education, health and housing decreased living conditions and increased both family poverty and child poverty (Zamfir 2002, 19-20). First of all, it was an unexpected transformation, therefore neither the families nor the state authorities had opportunity to calculate the social costs of this process. This lack of focus on social costs is clear in the economic deterministic preparations of the reform measures. Indeed, it is claimed that impoverishment in Romania had begun since 1980s (ASIS 2004, 11). In the late 1960s, the socialist economies began to suffer from structural crises in their economies, and in late 1970s they were affected from the worldwide oil crises. The economy had deteriorated in the last years of the previous system with increasing food scarcity and fall of living standards in addition to a largely obsolete industrial system. What the transition to a market economy changed is that opportunities and commodities have become more available while the inflation and unemployment prevented people from access to these commodities and services. The new system created winner and loser families, the pockets of poverty (relative poverty) rose hand in hand with the rise in economic growth. And most importantly the new system lacks the social protections provided by the communist state. In the communist period, full employment was an objective and most of the protections were about workfare. The minimum wage had been high enough for a decent life and social benefits were comprehensive together with pensions. High child allowances, free education and health care, highly subsidized and even free children's goods and services, extra allowances for mothers who had three or more children therefore could not work, highly subsidized housing rents were available (Zamfir 1997, 3). Cutbacks in these provisions in the transition period increased education costs of the children, and the children started to dropout the school to work on the streets. The previously controlled social divisions acquired a social and cultural aspect in the post-1990 period. The predominance of Roma children among the working street children indicate a positive relationship between social exclusion and inequality and working street children and child poverty, through which a connection with New Poverty and working street children might be established.

81

In addition to the factors relevant to poverty, the resources point to the family violence and disintegration as one of the main causes why children go to work on the street (Save the Children Romania 2003). Many children run away from the violence and conflict in the house and go to the street. Another aspect of family disorganization is that many families, who migrated to the cities during massive industrialization but lost their jobs in post-1990 period as different skills are required, and lost their homes with the cutbacks in subsidies, live on the street. Children in this category are the most vulnerable group (Save the Children Romania and IPEC 2003, 32). The increase of adult deaths and increasing rate of divorce increased the number of single parent households who live in severe poverty conditions and their children had to work on the street.

Demographic factors compose another causality in that the pronatal policies of the communist period resulted in high fertility. Contraception and abortion had been subject to strict controls and there were additional subsidies to the families with three or more children. The birth rate had fallen due to the beginning of economic hardships in early 1980s, but the state's pressures to increase birth rate had also increased with additional measures. This led to the poor families to have more children (Ghinararu 2004, 44-45). In the early 1990s, the child population over the total population was around 30%, and today the rate is around 25% (Ghinararu 2004, 45). Crowded families suffer from poverty more, the family budget available for each child to go to school is lower and thus the children go to work on the street.

This high birth rate combined with the hardships of unexpected transformation resulted in child abandonment (Ghinararu 2004, 45). Every year around 10.000 children are abandoned in maternity wards or in hospitals (Ghinararu 2004, 45). Some of these children have been either adopted (in many cases through international adoption) or sent to the institutions. Today, it is claimed that an important portion of children on the street are those who ran away from the residential care institutions because of disfavourable living conditions. These institutions had been established during the communist period in order to provide care to the children of poor families or the unwanted children who had to be delivered as a result of the lack of access to contraception and abortion (UNICEF Innocenti Research Center 1999, 19). After the transition, the birth rate decreased dramatically, however, the number of children in residential care increased due to the economic hardships and family disorganization. The mentality of the families that state has a responsibility to take care of the children remained the same in majority, thus, they keep sending their children to residential care despite bad living conditions.

Table 4.1 indicates the trend of institutionalization between 1997-2002 according to the statistics of National Authority for Child Protection and Adoption (Save the Children Romania and IPEC 2003, 12). According to the table, number of children in childcare institutions declines until 2000, however, due to the economic crises, there is a rapid increase in this number. Since 2000, this number tends to decline. Due to the policies targeting the children in the institutions, many children are taken from these places and many children are given to the supervision of substitute families. Therefore, there is a continuous increase in the figures between 1997 and 2002. In general, Table 4.1 indicates that number of children under protection in the institutions and substitute families have increased in this period.

Table 4.1 Institutionalization in Romania (1997-2002)

The reference moment	Number of children in childcare institutions	Number of children in substitute families	Total number of children protected in childcare institutions and in substitute families
June 1997	39.569	11.899	51.468
December 1998	38.597	17.044	55.641
December 1999	33.356	23.731	57.087
March 2000	33.600	25.433	59.033
December 2000	57.597	30.572	88.169
December 2001	49.965	37.553	87.518
February 2002	49.750	38.615	88.365

The decline in international adoptions increased institutionalization as these adoptions were taken under control because of the abuses relevant to these children and the failure of many attempts to find alternatives to these institutions ended up children in substitute care to go to the institutions (UNICEF Innocenti Research Center 1997, 14). These children mostly run away to the streets eventually. Therefore, the increase in institutionalization means an increase in the number of working street children.

The final reason is the family attitude towards children working. The institutionalization culture is one aspect of this fact in that they transfer their responsibilities to the state and the children eventually have to find other survival mechanisms (Westhof 1997, 4). In general, the parents' have favourable attitudes towards child labour (Save the Children Romania 2001, 33) and some of them are not aware of the dangers for their children. All these factors combined with the children's need to survive and/or desire to contribute to their families increase the number of children working on the street.

4.4.2 Scope of Working Street Children in Romania

In Romania, working street children are part of the new form of child labour. Therefore, just as done in the chapter on Turkey, a brief introduction to the child labour in Romania might be helpful to understand the scope of working children.

According to the *Survey on Children's Activity in Romania* conducted by National Institute of Statistics Romania in collaboration with ILO in 2003, number of children in 5-17 years is 3,866,438. Girls account for 48.9 % of the total, 48.7 % of the 5-9 year age group, 49.1 % of the 10-14 year age group are girls and 49.0 % of the 15-17 year age group are boys (Romania National Institute of Statistics 2003, 59). The number of children below 5 is estimated to be around 1 million (Romania National Institute of Statistics 2003, 55). The rate of children in economic activity is 4-5% according to statistics, however, the inconsistency with official figures and overview of poverty levels suggest that the percentage is high. Between 58.8-62.7% are boys and 35.3-32.3 % are girls in economically active children ((Romania National Institute of Statistics 2003, 55). 87.6% of these children are reported to work in agriculture, 6.8% in industry and construction and the remaining part work in service sector and other activities (Romania National Institute of Statistics 2003, 72). Inclusion of household activities and domestic work would arguably change this distribution however, it was not included in the official statistics as part of economic activity. The main reasons behind working are family poverty and parents' forcing the children to work or children's desire to contribute to the household budget.

In terms of working street children, in official statistics, the share of children on the street is generally stated to be around 1-2% of all economically active children (Ghinararu 2004, 16). In terms of gender dimension, more boys are working and living on the street compared with girls. One of the first studies state that 68% of working street children are boys, while the girls compose 32% (Save the Children Romania 2001, 25). For the children working on the street and returning to their houses at night, the gender differences is explained by the families reluctance to send the girls because of increased dangers, and that domestic work is more available for girls. For the children working and living on the streets, the girls tend to cope with family conflicts, violence and poverty as they fear street environment (Save the Children Romania and IPEC 2003, 17).

Most of the children are in the 12-13 age group although there are children from every age group (Save the Children Romania and IPEC 2003, 18). Families of these children are those migrated to the city during mass industrialization; they are low skilled, low educated, crowded and mostly unemployed families. The Roma population accounts for only 10% of total population in the country, however, almost half of the working street children are from Roma origin (Save the Children Romania and IPEC 2003, 18; Save the Children Romania 2001, 26). This finding not only affirms the relation between working street children and new poverty in Romania in that inequality and social exclusion play role in moving to severe poverty, but also points to an ethnic problem that might exacerbate in the future and threaten social cohesion in Romania. Poverty as the increase of extreme poverty among Roma community is the main reason, and both low education level and social disfavour towards Roma children prevent them to find jobs other than those on the street. Moreover, the cultural approach of the Roma community is claimed to state that children must work to contribute their families (IPEC 2002, 28). Low parental education might be another reason why they attach lower value to education and send to street to work in Roma communities (Ghinararu 2004, 63).

In terms of activities on the street, begging is the most common task (44%), car washing (17%) and peddling (15%) follow it (IPEC Fact Sheet 2005). Loading/unloading, household work, collection of waste products, stealing and prostitution are other occupations (Save the Children Romania 2001, 27). According to the study on working street children in Bucharest, children have been engaged in different tasks before they start performing these tasks on the street and some of the children perform more than one task at a time (Alexandrescu 2002, 27). The overwhelming activity being begging indicates that more girls and younger children are involved compared with other countries (IPEC Fact Sheet 2005). Children beg at crowded places by telling or singing their life stories and sometimes they are accompanied by younger brothers/sisters. Car washing involves washing the windscreens at intersections or washing the car entirely. Peddling is considered as one of the least dangerous activities performed by the

85

working street children in Romania. The items sold are cheap and the children sell items such as newspapers, car deodorants and city maps. Some children are engaged in peddling together with their families on the markets set up on the street. Teenagers and sometimes younger children work in loading and unloading in the places like supermarkets, construction sites and commercial areas. Household work is identified by the children as a second work which they perform when they are not begging or working on the street. Some children also do household work in the houses of people their families are in-debted. Finally, collection of waste products involves recyclable materials such as iron, glass and paper. This activity is reported to be performed by the *Roma* children (Alexandrescu 2002, 27-28). Time spent on waste collection depend on the season. Ethnicity is mentioned in the reports as the most important factor in determining the type of work (Alexandrescu 2002, x). However, *Roma* children are also engaged in other activities in some neighbourhoods. One report state that as children under 16 are not punished by the legal system for trading illegal goods, and some adults thus use the children for this kind of trade (Save the Children Romania 2002, 25). A police officer in Romania states that "What do children do? Small trade. They buy en gros and resell, sometimes even at double prices... The parent stands somewhere across the street with the bag and leaves the child two shampoos. When one is sold, he goes and gives him another one. I cannot pick up the parents...And I cannot intervene for the child, I can give him a fine and if he doesn't pay it, there's nothing I can do" (Save the Children Romania 2002, 25). In terms of stealing, drug dealing and prostitution, the two available reports on working street children in Romania – one on specifically Bucharest and the other one on Bucharest, Iaşi and Craiova- do not specifically refer to *Roma* children. However, a report on *Roma* children refer that *Roma* children are also engaged in these activities. A social assistant making research on *Roma* children in Bucharest explains that "The whole area of Livezi and Zabrauti is also dealing with drugs. Theft and robbery in order to obtain drugs for consumers. Others are only dealers. If you ask for a 'ball', it is 50,000-100,000. Children, parents, they are all drug dealers. Clients are also from the outside. Some of the clients were priests." (Save the Children Romania 2002, 26).

In terms of education, decreased quality and increased cost of education cause drop-outs and never-attendance (Alexandrescu 2002, 26). 62% of the children dropped out of school. Around 20% of the children had never gone to school (Save the Children Romania 2001, 31). Only one third of the children in Bucharest are reported to go to school while they are also working on the street. Existence of non-attendance schools and special schools for working street children in Romania might be claimed to contribute to school attendance among the children (Alexandrescu 2002, 26). On the other hand, high percentage of children who have never attended school compared with other countries might stem from *Roma* dominance among the working street children.

4.4.3 Response to the Problem of Working Street Children in Romania

The problem of working street children growing and becoming more visible and TV broadcasts about residential care institutions with degrading living conditions and thus indicating that more children will run away and start to live on the street; Romania decided to participate in the international struggle against child labour. Firstly, The National Action for the Prevention and the Elimination of Child Labour in Romania was established in 1999 under guidance of ILO, and the IPEC programme was initiated in 2000. The main intervention areas, targets and policies are identified in line with the general agenda of IPEC. Several research activities and surveys are conducted in line with IPEC's SIMPOC methodology, action programmes are carried out, direct support programmes, capacity building programmes and awareness raising action programmes are initiated (Ghinararu 2004, 11).

Working model of the institutions has been devised in line with IPEC as Romanian government is involved through the Memorandum of Understanding, National Steering Committee (NSC) leads the process with the advisory body of National Advisory Group on Child Labour. Special Child Labour Units (CLU) were also established together with National Authority for Child Protection and Adoption (NACPA), the Ministry of Labour, Social Solidarity and Family (MLSSF) and the Labour Inspectorate.

In terms of the international legal framework, in 1990, Romania ratified the UN Convention on the Rights of the Child. 28 September 1990. The ILO Minimum Age Convention (No. 138) was ratified in 1975, and in 2000, the country ratified ILO Convention on the Worst Forms of Child Labour No. 182.

In terms of the national legal framework, the Education Law (No. 84/1995, 28th of September, republished in 1999) extends compulsory education to 9 years, provides opportunities for uneducated children in that special classes might be established for those who have not completed the first four years of the compulsory education by the age of 14, evening, low-attendance or distance education classes might be organized for those who are more than two years older than the respective school age (Save the Children Romania 2001, 9-11).

There is no specific Child Code in Romania, therefore several treaties and articles refer to children mostly under the name of "minor" referring to people under 18 years. Ministry of Labour and Social Solidarity, the Ministry of Health and Family, the Ministry of Education and Research and the National Authority for Child Protection and Adoption are responsible for supervision of these responsibilities. The main provisions are in the Romanian Constitution, the Labour Code and the Family Code.

Article 39 of the Romanian Constitution prohibits forced labour, Article 45 prohibits exploitation or the employment of children in activities that are likely to harm their health or morals, or that endanger their lives or normal development, employment of children less than 15 years. Particularly Article 45 might be evaluated in the context of working street children.

Article 7 of the Labour Code allows the children at the age of 16 and who does not attend the school to work; moreover, children at 14 years of age can work in temporary jobs with suitable conditions to their development and knowledge with the approval of their families or legal representatives. According to this article, the employing enterprises have the responsibility to monitor whether they complete the compulsory education or not (Dutu 2002, 24-25).

Articles 161-162 limits harmful, difficult or hazardous works and night works to the age of 18. Article 112 limits working hours, and Article 118 governs the extension of working hours. Articles 84 and 160 sets rules for vocational training and apprenticeship. Article 7 of the Labour Code, states that "any individual aged 16 who does not attend school has a moral duty to perform work that is useful to society, that children aged 15 may be engaged in temporary work, whereas industrial work can only be performed starting at the age of 16, teenagers between 15 and 16 years of age can only be employed with the consent of their parents or legal guardians and only in work that is adequate for their physical development, their knowledge and their skills."

The Family Code warns the parents in Article 101 to consider the physical and moral development of the children when they would allow them to work, and requires a medical certificate to protect the children in the institutions (Dutu 2002, 28). Article 97 of the Family Code puts that the parental rights shall be exercised in the interest of the child.

Finally in the Penal Code, in Article 678, special provisions are introduced in the transition period relevant to trafficking in minors and use of children in pornography, exploitation through forced labour, slavery, prostitution, and activities violating fundamental human rights and freedoms.

A review of the legal framework reveals that Romania is eager to ratify the international conventions on child labour. The national framework has considerable in terms of protections provided to the child labourers which might be considered as a positive legacy from the communist period. However, the same legacy that work is perceived as both a right and a duty leads to a positive attitude towards child labour on behalf of both the state and the society as the society is not much aware that most of the new forms of child labour no longer mean skill investment. Table 4.2 summarizes the legal framework with regard to child labour (Dutu 2002, 28). According to the table, education for children below 14 years is compulsory, and they are not allowed to work outside the house. For the children older than 14, legal protection decreases to an extent, however, only light work is allowed and with written consent of the parents. For the children aged between 16 and 18, industrial work is allowed, yet, hazardous or hard work is banned.

Table 4.2 Legal Framework on Child Labour in Romania

14 years	15 years	16-18 years
Compulsory education	No compulsory education	No compulsory education
	Secondary education continued in vocational schools	Secondary education continued in vocational schools
	Written consent of parents for employment	No compulsory education
	Medical certificate	
Household, artistic and sporting activities only	Light work	Industrial work
		Banning of hard or hazardous work

89

With regard to the achievements and deficiencies in terms of working street children, the extension of compulsory education to 9 years in 1999 is very important. The causality between child labour and education has been investigated in many studies and it is known that extensions in compulsory education pulls many children back from the streets to the schools. In Romania, 1,826 children have been withdrawn from work or prevented to enter work through education or training opportunities or other services (legal assistance, counselling, health services, nutrition, uniforms, books and school supplies, stipends, other incentives) (ILO Romania 2005). At the end of 2002, the number of children living on the street is decreased from 2500 to 1500 (European Union and Government of Romania 2003, 22).

Conduct of researches in terms of working street children and the Roma children provide very significant data on which the policies might be modeled. The direct support programmes targeting the children, their families and the school is an effective model to be followed.

Alternatives are devised in order to replace the residential care institutions. Foster care and child homes are proposed as alternatives and the children are tried to be reintegrated to the families (Penton 1999, 136). The role of NGO, particularly Save the Children Romania is very important in this respect in that there are houses established for the children to have shelter such as Roxana House, which is specifically dealing with HIV infected children, a problem on the increase among working street children (Save the Children Romania 2005).

In terms of institutional structure, Romania adopted the IPEC model, however, the lack of data and the organizational problems with regard to the programmes, particularly due to the instability and political variance of the transition period, show that there are problems with enforcement (Dutu 2002, 7). Again, particularly due to the specific conditions of the country, there are legal inconsistencies between international conventions ratified and constitutional provisions with regard to child labour. For instance, the age limits of the constitution are in conflict with ILO Convention No.138.

As an achievement, inclusion of working street children into the provisions of the National Anti-Poverty and Social Inclusion Plan Concept (2002-2012) is an important step to include the problem in the mainstream policies against poverty eradication. Moreover, it consolidates the relation between struggle against worst forms of child labour and struggle against poverty.

90

Romania is a country whose economic and political system changed in 1990s. Since 1990, the country is trying to consolidate its political system and making reforms in order to adopt its economy to the market economy. Almost half of the Romanian population is claimed to live in poverty (Zamfir 2002, 7). In this context, participation in the IPEC framework, in my opinion, might be evaluated as positive development in terms of child labour and working street children. Since 1992, Romania has ratified 54 ILO Conventions, projects implemented for reorganization of employment, there are research activities on child labour, child trafficking, drug use and working street children, training activities and seminars are organized for trade unions. IPEC framework is based on the collaboration between governmental, civil society and international organizations in Romania, similar to the programme in Turkey.

The programme has similar shortcomings with Turkey. First of all, IPEC framework is a general programme and application of the projects which are designed according to the general international methodology is tried to be supported by studies on the country-specific factors and characteristics. The available reports are based on the research on *Roma* children, working street children in Bucharest, İaşi and Craiova and rural child labour in five cities of Romania. Studies exploring the specific causes that lead to emergence of working street children should be increased.

As mentioned in the chapter on Turkey, education dimension should be also emphasized. Compulsory education is extended to 10 years and a project is ongoing on the social protection measures for the children of the low income families to increase their access to education. However, I did not come across further studies on the education dimension. Research on the factors for school drop-outs, quality and cost of educations, attitude of the teachers and classmates, and school performance should be increased in order to provide a more developed basis for the policies. Particularly, education level of the *Roma* children is stated to be low (Save the Children Romania 2002, Zamfir 2002). Therefore, studies and policies on *Roma* children can be increased.

In Romania, likewise Turkey (Dayıoğlu and Hoşgör 2004, 46), NGOs lead the projects in IPEC framework. One particular NGO, Save the Children Romania (*Salvati Copiii)* seems the leading NGO involved in the projects on working street children and child labour, since the available reports on child labour within the IPEC programme are conducted by this organization. Increasing the number of NGOs involved in the IPEC framework might develop the approach of the studies and provide detailed and different information in my opinion. Involvement of NGOs is important in the Romanian case as the families and children might be uncomfortable with speaking to the researchers of the governmental institutions; however, in my opinion, collaboration between NGOs and governmental institutions should be increased and the need for social policies targeting children and their families on the state level should be emphasized. Otherwise, mainstreaming the efforts for eradicating child labour and working/living on the street might remain limited to certain districts in big cities and children at risk might not be prevented from working and/or living on the street.

In this chapter, working street children in Romania is discussed in terms of the causes, scope and dimensions in relation to New Poverty. The next chapter aims to make a comparison between Turkey and Romania on the so far discussed points of working street children.

CHAPTER V

WORKING STREET CHILDREN AND CHILD POVERTY IN TURKEY AND
ROMANIA IN THE CONTEXT OF NEW POVERTY

5.1 Introduction

A comparative analysis of Turkey and Romania in terms of working street children and child poverty particularly in the context of New Poverty has certain difficulties. Notable among these difficulties is the problem with qualitative and quantitative data. The notion of working street children is a relatively new social problem dating back to late 1980s and early 1990s. The fact that studies on child labour and working street children dates back to earlier years and Turkey has initiated its institutional struggle with worst forms of child labour earlier provides access to more detailed and concrete data. On the other hand, Romania joined IPEC in 2000, 8 years later than Turkey's participation; and there are fewer studies compared with Turkey.

Collection of data on working street children is difficult in that they are difficult to be reached and their families are reluctant to give information or accept that it is a problem. Moreover, while institutional dimension on the one hand means more research and more policy, on the other hand, it puts normative limitations. Since both countries have problems with political stability and as they are candidates to the European Union membership, it might be claimed that they prefer emphasizing certain points at the expense of others. For instance, *Roma* problem is a political problem in Romania, the welfare and living conditions of whom are stated among the political criteria for the accession of Romania to the European Union (European Commission Directorate General Enlargement Information and Interinstitutional Relation 2003, 8-9). There are statements, particularly in the official documents that despite the high percentage of *Roma* children among working street children in Romania, the ethnic dimension is not a cause but rather an effect (Ghinararu 2004, 62). As a result, it is very difficult to evaluate social exclusion dimension of working street children and of child labour in Romania. With regard to Turkey, there are ethnic references in the form of reference to the social unrest in the Eastern and

Southeastern regions the subsequent migration following the social and political unrest (Akşit et al. 2001, x). There is also direct ethnic reference in terms of the origin of the children (Kahveci et al. 1996, 41). However, other than the statements of some working street children that some people mistreat them because of their ethnic origin (Kahveci et al. 1996, 65), ethnic dimension is not explored as a factor or characteristic. The European Union documents do not also discuss the ethnic dimension in detail; instead, there is a reference in terms of the problem of giving children names other than Turkish (European Commission Directorate General Enlargement Information and Interinstitutional Relation 2003, 88).

In addition to these problems, in the documents originating from Romania, it is difficult to determine the scope of previous regime's legacy as a problematic factor behind working street children. In the official documents, problems with regard to inadequate transition to market economy, rather than the social consequences of such an economic-political system is put forward. The fact that transition to a market economy and the restructuring that followed had high social costs is acknowledged (Romanian National Institute of Statistics 2003, 43), nevertheless, interruptions in reform process and interrupted economic growth is more emphasized. The structural problems in relation to the previous system and collapse of the system is frequently mentioned, however it is observed that these statements can be taken as ideological standing points since the claims are not explained with concrete data and examples.

In terms of the scope and aspects of working street children, some points are left unanswered. Especially there is lack of data with regard to illicit activities such as prostitution and drug dealing. This is first and foremost, because these activities are less visible, and they constitute a serious criminal activity. It also stems from the fact that the studies in majority focus on working street children in general. And majority of the children have contacts with the families, in other words they are not children of the street, therefore, other characteristics might come forward as they are more common among the subjects of the research. The approach in terms of prostitution is more open in the Romanian documents (IPEC 2003, 26). There are references to the children on the street that some of them are engaged in prostitution (Save the Children Romania 2003, 26). Indeed, it is reported that 35% of the girls state that they are engaged in prostitution and 20% of the boys stated that they are abused by adult pedophiles at least once (Alexandrescu 2002, 7). This dimension is not much discussed in the corresponding IPEC research in Turkey, which might be explained by the cultural dynamics or lower number of

94

street children. The cultural context by not approving working of the girls on the street after puberty protects the children in a way in Turkey (Akşit et al. 2001, 68). And in my opinion, it also prevents discussion of matters such as prostitution especially in the studies involving governmental institutions. Küntay and Erginsoy's study (2005) on girls engaged in prostitution stands out the most detailed study on prostitution of underage girls in Turkey. However, this study does not cover working street children in general, rather focus on the specific group of girls engaged in prostitution in İstanbul. In terms of boys, it is stated in the study on shoe-shine boys in İzmir that boys are "threatened by people intent on sexual abuse" (Kahveci et al. 1996, 57).

This chapter is devoted to the comparative analysis of Turkey and Romania in terms of working street children. The scope and dimensions of child poverty will be compared in advance, trying to establish a link between the course of child poverty in both countries and working street children. The main aim behind this comparison is to gain insight into the transformation of poverty and the change in the way children are affected from poverty through the comparison of Turkey and Romania and to explore emergence of working street children in both countries at similar time periods. The final part of the chapter is devoted to possible outcomes.

Before going through the specific aspects, it should be noted that international documents involving general evaluations in terms of the progress made by countries and the situation of people have controversial comments on Turkey and Romania. Given that Turkey was established in 1923 and Romania is a transition country trying to establish even basic institutions, one might claim that Turkey progressed relatively more than Romania. During my visit to Romania in April 2004, the living conditions and access to the basic services, even in the capital city, had seemed to me far away from comparison. The devastating effects of a political conflict and transition were visible. Nevertheless, most of the international documents rank Romania above Turkey in terms of poverty, human development and child poverty, particularly due to the positive influences of social policies implemented during the previous era. This point is further discussed in the next part.

5.2 Children in Turkey and Romania

The latest IPEC reports regarding the progress in terms of children ranks Romania above Turkey, at 47 and 48 respectively (IPEC 2005, 41). This evaluation is made on the basis of achievements relative to the targets of legal framework, policies and programmes, mainstreaming and data collection in the context of child labour. Turkey is in the Time Bound Programme, which means that it can solve its problems on its own at a given time interval, might be conflicting with this ranking; however, this evaluation is specific to the 2004-2005 period.

The *State of the World's Children Report* is more specific in terms of the situation of children. In terms of the under five mortality rates, Turkey ranks at 79, Romania being 120th among almost 200 countries (UNICEF 2004, 105). This means that many more children die in Turkey before the age of 5. This ranking might be linked to the high regional discrepancy in Turkey. The health facilities in the eastern and southeastern regions are inadequate in terms of availability of physicians/medical personnel and medical equipment. According to the State Planning Organization data, Eastern and Southeastern Anatolia have considerably lower numbers of physicians/medical personnel compared with Marmara, Aegean and Mediterranean regions (SPO 2002). Part of annual state budget devoted to the health expenditures is around 3-4 % of the national income, lagging quite behind the European countries (Labour Union of Health and Social Service Workers 2006). The average level of health spending between European Union member states is around 8% of the national income according to 1998 figures (Emmerson et al. 2002, 2). Unequal allocation of available resources added, the regional living conditions differ. It is also noted that the political unrest in these regions, targeting of medical facilities during the conflict further worsens the health conditions (Karabulut and Emsen 2003, 21). On the other hand, despite the decreases in the health expenditures, the established health sector might be considered as an advantaged position in Romania. This point is further strengthened by looking at the change between 1960 and 2003 in that under five mortality rate per 1000 children dropped from 82 to 20 in Romania, while decreasing from 219 to 39 per 1000 children in Turkey. Therefore, it might be claimed that pre-1990 investments in Romania on health provides such a situation, and this interpretation also reveals that Turkey made more progress in a shorter time period.

The two countries are very different in population size, Turkey's population is above 70 million, while Romania is hardly above 20 million. Therefore, a comparison in terms of rates might be more explanatory rather than figures. The same report states that life expectancy at birth is similar in both countries by year 2003 (UNICEF 2004, 108). The comparison of adult literacy rates, 98% in Romania and 85% in Turkey, follows the line of argument that previous policies prevail. The share of household incomes between 1992 and 2002 is a little bit complicating. The lowest 40% gets 21% of the GDP while highest 20% has 38% in Romania. In Turkey, the lowest 40% gets 17%, while the highest 20% of the population has 47% (UNICEF 2004, 108). Thus, it might be claimed that pre-1990 social situation is changed in favour of increasing inequality in Romania between 1992-2002, particularly due to the cutbacks in social provisions and subsidies in transition; and Turkey still has higher inequality notably due to the social policy changes in late 1980s in line with structural adjustment programmes.

In the period between 1960 and 2003, statistics indicate that life expectancy in Turkey increased from 56 to 71, while it increased from 68 to the same level in 2003 in Romania (UNICEF 2004, 128). From this data, and the previous data given above, it might be concluded that in terms of living standards, Turkey made continuous and rapid progress, while Romania is losing pace and the living conditions which had been established previously is deteriorating in the transition process. According to the economic indicators, GNP per capita is higher in Turkey. While the GNP per capita average annual growth rates were similar until 1990s, after 1990s the rates dropped in both countries, the drop in Romania being almost twice (UNICEF 2004, 132). The economic indicators might be interpreted in that in 1990s, integration to the international political economy had social costs for both countries as both countries have begun to apply neo-liberal policies, however, the implications are tougher in Romania due to the transition. In the first chapter, it was discussed that global dynamics and the neo-liberal paradigm have transformed poverty particularly in the developing world, relative poverty and inequality increased and this new form of poverty is called as *New Poverty*. In Turkey and Romania, a similar process has taken place. Some sections of the society in both countries are impoverished and their income dropped drastically due to inflation, unemployment, lower real wages and increased costs of education, housing and health; all this in turn pulling down the GNP per capita rates.

In terms of demography, fertility rate in Turkey is above Romania in general. At the beginning of 1990s, fertility rate in Turkey was 6.4 and it dropped to 2.4 in 2003. In Romania, the fertility rate was 2.3 in 1990 dropping to 1.3 in 2003 (UNICEF 2004, 144). Both countries have been subsistent type agricultural economies until 1970s. In Romania, the pro-natal policies of particularly Caucescu regime resulted in an artificial increase in 1970s despite the tendency to decrease as the economic situation was on decline. In the beginning of 1990s, the fertility rate dropped suddenly as the transition led to an initial collapse in economy. The decline in the fertility rate still goes on as the economic situation keeps being severe. On the other hand, in Turkey, the low education level in eastern and southeastern regions and the persistence of agricultural economy increases the regional fertility level, resulting in a higher rate countrywide. With the increase in literacy and extension of compulsory education, the fertility rate is declining, however, it might still be considered as higher than the European average (GAP 2000).

Generally speaking in terms of human development levels, the research indicate that despite the difficulties experienced by Romania, particularly after 1990s, given the criteria of life expectancy, educational attainment and adjusted real income, Romania has better conditions. The human development index established by UNDP in 2005, ranks Turkey at 94 and Romania at 64. Both Turkey and Romania are considered by UNDP as medium income countries ranking in the medium human development category (UNDP 2005, 379-380). Another report states that inequality, measured on the basis of income and expenditure in the specific framework of consumption, is also higher in Turkey (World Bank 2006a, 41).

The economic problems of the two countries have distinct historical context. Romania was out of the market economy until 1989. The dissolution of the Eastern Bloc is resulted in economic and political difficulties for the country. The policies towards a market economy on the cutting of state social services accompanied by privatization, corruption, inflation, unemployment and poverty shook the Romanian society. In contradiction, Turkey, as early as its foundation in 1923 as a republic, followed the Western path. The First World War had exhausted the society and resources, and the establishment of a new state after the Ottoman Empire has given birth to a Republic with economic difficulties.

Although Turkey made a remarkable advance up until 1950s, its fragile economy could not overcome the Depression of 1930s and Second World War. Coming to early 1980s, Turkey's budget deficit rocketed and indebting to IMF and World Bank resulted in a change of economic policy towards export-oriented growth (Ecevit 1998). The structural adjustment programmes has foreseen privatization, shrinking of state, increasing of export, creating an unorganized labour force, and other items for the working of a free market economy. The consequences of these policies on the society became clear in 1990s, and Turkey and Romania's roads have merged since 1990s in terms of their common socio-economic problems involving "children working and/or living in the streets".

In their relations with EU, the issue of children of Turkey and Romania are mentioned in official EU documents. These statements are usually indirect and are included in the Chapter 13: Employment and Social Policy and Chapter 18: Education and Training of Enlargement Guide documents and in the Strategy Papers. The progress in terms of the situation of children is also referred in the Annual Reports. The difficulty with doing research and identifying the exact situation is caused by the fact that working street children is a comparably recent issue and is usually invisible (UNICEF 2005, 40).

Particularly speaking, EU states in the 1999 Commission Report under the light of criteria of Helsinki Summit that priority to the crisis in childcare institutions in Romania is among the prerequisites of the political criteria (European Parliament Resolution 2000, Article 49). In the Article 49 of European Parliament on Romania's Application to EU (2001), textual alterations with respect to abandoned child in Romania's incorporation to the UN Convention of the Rights of the Child, abuse and neglect of children in state institutions, the growth of street children problem and child trafficking, international adoption, and child protection are criticized.

2003 Regular Report on Romania expresses the PHARE support to child welfare and conditions of *Roma* as an integrated part of political criteria together with awareness raising of children's rights (European Commission Directorate General Enlargement Information and Interinstitutional Relation 2003, 8-9). The problems with the children in orphanages, child protection, decreasing the number of children in state residential care, disabled children, child trafficking, education of the children of the migrants are expressed and immediate action is urged in the following chapters of the same document. In the Roadmap (Commission of the European Communities 2002, 36) and Enlargement Guide (European Commission 2004, 59), under the Chapter 18: Education and Training, education of children of the migrants is again mentioned.

The European Union criticism towards Turkey has some distinct points in the documents. The child trafficking, abandoned children, education of the children of the migrants, international adoption and children in the state residential care are not mentioned with regards to the EU papers on Turkey. However, the criticism about child labour is more to the point since available data is much more comprehensive with an additional 8 years of research by IPEC in Turkey. Moreover, the situation of children are not claimed to be an integral part of the political criteria for accession in Turkish case. Possible reasons for this might be Turkey's problems about children (for instance international trafficking and adoption of children are not much common, at least is not widespread enough so as to be visible.) might be considered by EU as relatively minor. Another possible reason might be the fact that the combat with child labour is more institutionalized and government policy is on application since IPEC activities in Turkey began earlier. Therefore, progress with the fight against child labour might have been considered.

In the 2003 Report on Turkey, it is mentioned that child labour persists, names other than Turkish origin given to the children is still subject to state opposition and problem with "honour killings" of out-of-wedlock children continues. It is also criticized that Turkey placed reservation on Article 13, paragraphs 3 and 4 on choosing of a non-public school and religious and moral education for children, services for the children and women of internally displaced children are inadequate, street children are exposed to physical, sexual and drug abuse together with police brutality (European Commission Directorate General Enlargement Information and Interinstitutional Relation 2003). Moreover, the Turkish Employment Organization and Child Bureau are considered to be deficient (Ghinararu 2004, 88).

Turkey is one of the first countries which signed an agreement with IPEC and started its National Action Plan for the Prevention and the Elimination of Child Labour as early as 1992, and now its in Time Bound Programme. This is very important since it indicates that Turkey made progress in its struggle against child labour and in particular in working street children and now it can solve its problems on its own. Romania on the other hand, started its programme in 2000, eight years later than Turkey. While, its latecoming might be compared with Turkey as a disadvantaged position at first sight, however, given the country's market economy experience and a Western type of living has begun in early 1990s, it is notable that Romania is the first country in Central and Eastern Europe to participate in IPEC programme (Ghinararu 2004, 11).

5.3 Working Street Children in Turkey and Romania

There is a number of research on working street children in both countries. In this part, the basic research reports prepared in both countries under the supervision of IPEC is preferred. This is for two reasons. First of all, the terminology on children working and/or living on the street differs in other researches. The use of IPEC methodology provides a form of unity in terms of what is a child and how come a child becomes a child working on the street and child of the street. On the basis of such a terminologic unity, it is more feasible to make a comparison.

Secondly, the documents prepared overwhelmingly by the governmental institutions might lack objectivity in determining the causes and scope of working street children. Involvement of universities and civil society establishments might increase objectivity of the studies.

The main reference in this part for comparing the situation of working street children in Turkey and Romania is to three basic documents: For Turkey, Working Street Children in Three Metropolitan Cities: A Rapid Assessment (Akşit et al. 2001); and for Romania, Baseline Surveys on Working Street Children: Bucharest, İaşi and Craiova (Save the Children Romania and IPEC 2003) and Romania, Working Street Children in Bucharest: A Rapid Assessment (Alexandrescu 2002) Review of these documents are not at the expense of other studies, but rather it is an attempt for establishing a ground for comparison on the basis of these documents, which are conducted in similar methodological perspectives.

5.3.1 Comparison of Dynamics behind the Emergence of Working Street Children in Turkey and Romania

A comparison of the dynamics behind emergence and increase of working street children in Turkey and Romania indicates similar patterns at first sight, which then differentiate going into the details.

In the studies conducted for working street children in Turkey, poverty and migration are identified as the main factors (Atauz 1990a, Karatay 2000b, Kahveci et al. 1996, Akşit et al. 2001). Turkey has experienced rapid urbanization since 1950s due to the modernization in agriculture and increased work opportunities in the cities. These early migrants had followed a line of chain migration in which the first comers had established community networks that would help the late comers to find a house and a job. In this period of migration, according to Atauz, street children was not as "overt" as the situation today (Atauz 1989a, 37). In late 1980s and early 1990s, Turkey was faced with a new wave of migration. However, the cities, which are subject to another wave of migration, did not have the enough conditions to meet the needs of the migrant families (Atauz 1990a, 5). This new wave of migration has not only a "rural to urban" aspect because of the decreased opportunities in the countryside, but also has an "urban to urban" dimension. The armed conflict between state forces and PKK undermined the economic structure of the eastern and southeastern regions and resulted in a social and political unrest combined with the economic difficulties. This conflict, particularly effective between 1984 and 2000 – reemerged recently- inflicted a continuous flow of migration from the East to the West (Akşit et al. 2001, x). Karatay et al. (2000a, 432) refers to the new migration flow as "extraordinary migration" since migration flows in other countries occur usually for economic reasons. These new migrants had not effective available networks in the city not only because it was a rapid and sudden migration which did not leave time for cost-benefit calculation, but also because the material resources available in the city was disappearing due to de-*gecekondu*ization in particular (Buğra and Keyder 2003, 23), and decreased job opportunities available for low-skilled and low-educated migrants.

Cutbacks in free public services and increased cost of education, health and housing deteriorated the living conditions of them as migrants. Most of the male household leaders could not find a permanent job, as now the service sector jobs requiring high education and skills are increasing at the expense of others. Mothers, usually not speaking Turkish at all (Akşit et al. 2001, 63) and under the strict cultural constraints could not even work. High fertility rates added, these families tended to find survival strategies based on the labour of the children. Child labour is not a new notion in the Turkish society. The children of the poor families usually work on the family farms in the countryside, and in small workshops in the cities in order to contribute to the family budget and learn skills for their future occupations. Most of these children had been going to school and working in their spare time. As these kinds of jobs available for children disappeared by and large since 1990s, and as the education costs increased and family income decreased, the children of these families began to work on the streets. The deficiencies of the legal system in terms of -may be not child labour, but- working street children contribute to the increase in the number of children working on the street.

In the Romanian case, poverty of the families is also stated as the main factor (Save the Children Romania and IPEC 2003, 6; Alexandrescu 2002, x). The end of the communist regime and transition to a market economy accompanying the political integration to the Western world caused structural problems. Unemployment and inflation rose suddenly. The country has achieved economic growth for certain periods, however, interruptions occurred in the economy. The official documents state that interruptions rather than transition to a market economy caused such an increase in poverty. Moreover, they claim that incidence of child labour tend to decrease

102

in the period when the market reforms are executed more efficiently (Ghinararu 2004, 15). On the other hand, even the official documents accept that the decrease in social services and subsidies in particular, and liberalization in general increased relative poverty, for certain disadvantaged sections of the Romanian society (Ghinararu 2004, 15).

The second dynamic in Romania is pointed out as family disorganization. The economic difficulties in Romania increased extreme poverty comprising almost 40% of the society according to the 2002 statistics. Alcohol and drug addiction, crime rate and family violence increased leading to divorce. The social cost of this transition period increased number of single parent households. Increased unemployment and decreased incomes on behalf of the male household leaders might be claimed to lead to increased alcohol and drug usage and violence in the families. Increased family disorganization have resulted in many families that children are left with one parent to supervise them. A child interviewed in a report states that "... we were very young, he killed her on December the 14th 1991, I was 7 years old at that time. I spent a whole year in courts because my father wouldn't confess to killing my mother" (Alexandrescu 2002, 53). One fourth of the children in the research on working street children in Bucharest are living in a single parent household (Alexandrescu 2002, 35). And majority of these children are headed by mothers according to the same report. Fathers might be deceased, have left the family or be in prison.

Given that many of these families are headed by women who traditionally have more difficulty in finding a job and earning high income, even in a country with socialist history, puts these families in worse economic conditions. Children of these families are at the highest risk of going to the streets to contribute to the family budget. Those living in families with violence are also at stake as many of the children interviewed on the street state that they run away from the violence and prefer the street to the house (Save the Children Romania and IPEC 2003, 17). The low level of education and high fertility among the migrants which migrated to the city during massive industrialization before 1990, and increased cost of education doubling the relative cost of children when they are not working, and finally the deficiencies of the legal system are the other factors.

Comparing the causes why children work on the street in Turkey and Romania unite the two countries around 1990 despite their distinct histories. Turkey was already a Western-oriented country since its establishment; however, its economy was a kind of hybrid economy based on the general rules of market economy but under state regulation. Romania was a centrally planned command economy quite distant from the market economy despite some measures applied after 1970s.

Turkey transcended to the full-fledged market economy around 1980s because of the structural adjustment programmes and its high debt to the international monetary organizations, and Romania transformed its economy to market economy in 1990s because of a political transition inevitably requiring a market-type of economy. Prevalence of neo-liberal paradigm necessitating a structural transformation in the economy along with a decreased role of state (and withdrawal and/or limitation of public expenditures) helps us to explain both why these two very different countries had the same problem and why child poverty (and poverty) which existed even before 1990s resulted in working street children. To be more specific, poverty itself has transformed to a new kind of poverty in 1990s in which relative poverty and social exclusion dimensions are prominent.

The change in the political economic system globally and in the individual countries, on the one hand increased the need for the children to work and deprived these children from other options in the economy, for instance working in the industry. The family characteristics of the children in both countries reveal that the education level and skills of the parents added to the cultural disadvantaged position puts them in such a position that, no matter which country it is, they eventually go to the street. Thus, shortly we can answer to the question why Romania and Turkey have working street children as a worst form of child labour that the global dynamics and neo-liberal paradigm affect the internal dynamics of political and economic system. Coming to the question why child poverty resulted in working street children overwhelmingly, not another form of child labour -this is not to ignore other existing types of child labour, but as a prevailing type- is linked to the same explanation. Transformation of poverty also transformed child poverty.

Generally speaking, I argue that absolute poverty rates decreased in many countries including Turkey and Romania, however, child poverty has kept increasing even in the wealthy Western European countries and the US, as relative poverty was on the rise and children are the principal beneficiaries of the welfare provisions. Thus, withdrawal of state subsidies to especially education and health in line with neo-liberal discourse in both countries transformed the situation of children. The living standards of the children of poor families (in the New Poverty sense) decreased dramatically, the subsidies have been cutback, the opportunities to work together with school decreased both because of the increased educational costs and decrease of such jobs along with the transformation of the economic system.

In terms of differences, migration has different implication in two countries on children. In Turkey, economic difficulties and social unrest have caused migration to the big cities. The bulk of working street children belong to these newly migrated families since 1990. Particularly in Istanbul, children are coming from the recently migrated families (Atauz 1990b, 9). In Romania, there was a similar rural to urban migration in 1970s due to the second wave of industrialization. As most of these families suffer from unemployment in post-1990 period, some of the children of these families started to work on the street. However, the economic difficulty in the transition period gave way to a kind of reverse migration from the urban areas to the countryside (Ghinararu 2004, 42). As traditionally child involvement in the economic activity is higher in agriculture, child labour increased overall in Romania, this pattern of new migration flow means that those families and their children who had to stay in the cities despite severe economic conditions are mostly socially excluded groups to which street has become the sole option.

Emergence of homeless families in Romania in the transition period is another distinction (Zamfir 2002, 14). The higher rate of extreme poverty and maybe cultural dynamics led to the emergence of street families, higher rate of children living on the street among the working street children in Romania might be explained in relation to this dynamic.

There is also a causality in Romania due to institutionalization and child abandonment (Zamfir 2002, 16). During the communist period, high fertility policies were enforced on behalf of the government. On the other hand, as the prevailing paradigm in that period was that the welfare issues were under the responsibility of the state. Many institutions had been opened for the children whose families stated officially that they could not afford the expenses (Romanian Ministry of Labour, 2001). When the old system broke down, these institutions could no longer be financed, violence and abuse in these institutions made the children living at these institutions to run away to the street. A part of children working and living on the street come from these institutions (Government of Romania National Authority for Child Protection and Adoption 2003). The SHÇEK institutions might be thought in the first instance as a corresponding theme, however, the continuation of the policies supporting these institutions in Turkey might have

105

prevented such a consequence. However, children are also abandoned in Turkey (Kurt 2001) and there are over 700.000 children who do not have any families or relatives (Çocuk Vakfı 2000). In terms of the living conditions, some girls interviewed in Küntay and Erginsoy's research that they were staying in the institutions under the supervision of SHÇEK, however, they ran away together with a group of girls (Küntay and Erginsoy 2005, 180). The reasons for running away might be related to the desire to see the life outside and entertain; however, it might also related to the conditions at these centers. Some of the girls which later engaged in prostitution still do not want to go back as well. Economic difficulties, family disorganization and neglicence of the parents are identified as the causes for incidence of child abandonment in Turkey (Çocuk Vakfı 2000).

In Romania, many families, particularly single parent families, abandoned their children in Romania due to the economic difficulties and the understanding that state should find a mechanism to deal with. These two points added to the homeless families provide a difference between Turkey and Romania in that number of children also living on the street might be higher in Romania.

Finally, ethnicity provides another distinction point. Existence of *Roma* population in Romania puts forward the causality between social exclusion and working street children. The official statistics indicate that only 1% of the population is *Roma*. However, an almost 80% of working street children are from *Roma* origin. 87% of *Roma* population is claimed to be suffering from poverty (Zamfir 2002, 15). *Roma* population is more vulnerable to poverty in Romania because they have lower education and lack necessary marketable skills, higher fertility and they are exposed to discrimination (Zamfir 2002, 15). The *Roma* population is overwhelmingly out of the social security system and most of the children do not attend school due to cultural dynamics and the cost of education, finally going to the streets. This point strengthens the relation between New Poverty and working street children through social exclusion. Existence of *Roma* children working on the street is also valid for Turkey, however, no specific study is conducted for *Roma* working street children in Turkey, and the available studies refer to Kurdish and Arabic speaking families rather than the gypsies (Kahveci et al. 65). In Atauz's study, it is mentioned that girls living on the street only exist among beggars and gypsies (Atauz 1989a, 39). Instead, children from Kurdish and Arabic speaking families seem to be higher in number (Kahveci et al. 1996, Akşit et al. 2001). Depending on the statements of some children that the other children do not play with them as they are Kurds, and some fathers' complaints of unemployment on ethnic background, social exclusion dimension might be argued to be valid also in the case of Turkey. According to the study, one father says "The reason two of my children are shoe-shine boys is that people from Western Turkey don't employ us (Kurds) very easily." (Kahveci et al. 1996, 69). Although other studies do not mention discrimination on ethnic grounds, incidence of social exclusion in these cases might also imply a causality between New Poverty and working street children in Turkey. Why child poverty has led to emergence of working street children after 1990s in both countries, and why the two countries have a similar problem despite their distinctions have another answer here in that the global rise of social exclusion as a determinant in poverty is relevant.

5.3.2 Scope of Working Street Children in Turkey and Romania

Practically speaking, it is very difficult to compare working street children in Turkey and Romania in terms of numbers, as it is difficult to reach these children. Available figures basically rely on estimations. It is also the case that population in both countries is very different in terms of amount, therefore, a comparison based on figures might not provide correct comparison.

In both countries, most of the children working on the street are between 7-17 age group, nevertheless, there are also children below 7. In Turkey, the age median is 12; most of the children being between 9 and 11 years (Akşit et al. 2001, 35; Kahveci et al. 1996, 39; Karatay et al. 2000a, 447). In Romania, majority of the children are between 12-14 years of age, median again being 12 (Save the Children Romania and IPEC 2003, 17). This result might be interpreted as working street children cover almost all groups of age below 18, younger children are probably used in occupations such as begging and selling petty commodities in order to arouse sympathy, older children engaging in more dangerous jobs such as garbage collection. Age is a determinant of the type of job the child is engaged with, in both countries (Altıntaş 2003, 139-140; IPEC 2005). Tasks such as selling petty commodities are usually performed by younger children (Karatay et al. 2000a, 447).

In terms of gender, studies by Atauz (1990a, 1990b, 1996), Kahveci et al. (1996), Karatay et al. (2000a, 2000b) and Akşit et al. (2001) in Turkey; Alexandrescu (2002), Save the Children Romania and IPEC (2003) in Romania indicate that more boys are involved in street work in both countries. Girls either do domestic work, and if they sent to the street, they are usually below puberty age. This situation is also reflected on average age between children in the sense that average age of the boys is higher than the girls (Karatay et al. 2000b, 474). The gender difference is more visible in the Turkish case. While around 70% of the children are boys in Romania, this rate might be as high as 90% in Turkey. Therefore, a cultural aspect might be mentioned in this case. It is also to be noted that girls are claimed to be more patient in the case of family violence, therefore they might not run to the street to work in Romania (Save the Children Romania and IPEC 2003, 17). Given that family disorganization is identified as the second major cause, this aspect differentiates the Romanian case from the Turkish case. However, this distinction might also be relevant to the fact majority of working street children in Turkey are working on the street, only a small percentage of them are living on the street (Atauz 1989a, 37; Atauz 1996, 468). While the studies focusing specifically on the children living on the street also refer to the incidence of family disorganization in Turkey (Atauz 1996, Karatay et al. 2000b; Küntay and Erginsoy 2005), other studies made on the children working on the street (both children on the street and children of the street) state that family disorganization is not a significant factor (Akşit et al. 2001, Altıntaş 2003, Karatay et al. 2000a). The causes that lead to family disorganization in Turkey is explained as "loss of self-esteem, alcoholic parenthood (usually father), mistreatment of the children as a result of the frustration involved" which might result from increased unemployment and poverty in the households (Atauz 1989a, 40). In the Romanian case, existence of street families and homeless families in addition to street children might have strengthened family disorganization as a factor. As most of the research conducted are rapid appraisals, it might also be claimed that most visible and striking aspects are put forward due to time and money constraints. Thus, certain points might not have taken place in the studies.

The families of the children in Turkey are usually from the eastern and southeastern regions recently migrated to the Western cities. Mothers are usually illiterate, usually housewives, and spoke Kurdish in majority. Fathers are usually unemployed or underemployed in informal sector. Fathers usually speak Turkish, however, their education level is very low (Akşit et al. 2001, 62-63; Kahveci et al. 1996, 44; Karatay et al. 2000b, 513). In Romania, most of the children originate from northeastern Romania. The profile of working street children in Romania becomes dual in terms of family characteristics. For *Roma* children, their families usually cannot find in formal jobs due to the increased discriminatory behaviour and the parents are usually employed in informal sector if they can find a job. They speak *Roma* language, and parental education level is low. Lack of language skills and low education level might be interpreted as they live in deprivation (Save the Children Romania 2002, 38). For the rest of the children, parental literacy rate is higher due to the previous regimes free and compulsory education policy. However, parental unemployment is very common again. Most of the families had come to Bucharest and other big cities in the communist period due to forced urbanization and industrialization. In both countries, the families are crowded.

In terms of education, majority of the children in Turkey somewhat attend the school, only 13% of the children never attended the school, however school dropout rates are high being around 25% (Gündüz Hoşgör 2004, 6; Kahveci et al. 1996, 39). Discriminatory treatment of the teachers and other students, increased cost and decreased quality of education; crowded classes, lack of personal attention and difficulty with the access education are the factors behind high school dropout rates. The extension of compulsory education to 8 years contributed to increase of the educational level of those still working on the street, and decreased number of children that work. The IPEC fact sheet on working street children (2004) indicates that 62% of the children working on the street had dropped out the school. The same fact sheet states that 38% of the working street children in Turkey either dropped out the school or never attended; this might be considered to indicate educational discrepancy between the countries. Decreased public access to education and decreased importance devoted to the education due to decreased budgetary sources, increased poverty and education costs in the transition process are behind the school drop outs (Zamfir 2002). The fact that most of the children are from *Roma* origin, given that they have the lowest school attainment rates in Romania, might also be responsible for this difference, and it can be claimed that the children in Romania still have higher educational level. In both countries, the children interviewed state that education is important to them and they would like to go to school (Akşit et al. 2001, 79; Karatay et al. 2000b, 529; Alexandrescu 2002, xi). This means that high rate of school dropouts is not a personal choice on behalf of children, but rather an outcome of the economic conditions.

Most common occupation of the children in Turkey are shoe-polishing, selling petty commodities, scavenging, car cleaning, loading and unloading, garbage collection and separation. Although, there might be different according to location (Karatay et al. 2000b, 495), in general, shoe polishing is the most common activity.

On the other hand, in Romania, the most common task of earning is begging. A study on Romania claims that this finding mean that more girls and younger children are on the street compared with other countries (IPEC Fact Sheet 2005). Lower rate of begging in Turkey might be depended on cultural attitude to begging in general. Incidence of begging is mentioned mostly in the case of children living on the street (Zeytinoğlu 1989, 16; Atauz 1989a, 39). Therefore, this situation might also stem from the fact only a small percentage of working street children is living on the street. Other jobs in Romania are car washing, garbage collection, loading and unloading, household work, selling petty commodities, stealing and prostitution. In Turkey, age (younger children might be thought to be involved more in begging and selling petty commodities), gender (girls almost in all are engaged in selling, they are not seen in other jobs such as shoe polishing or car washing) and origin city (Altıntaş 2003, 141).

Persistence of migration as a factor might be claimed to lead to grouping of children from the same origin city in similar jobs just as the adult migrants do (Altıntaş 2003, 141). Whether they live on the street or not (majority of the garbage collectors permanently live on the street) determine the type of the work. In Romania, ethnicity is mentioned to be the most important factor in the type of the work. Most of the children in car washing and garbage collection, while Romanian children are more engaged in begging and household jobs (Alexandrescu 2002, x). Despite the social label on the *Roma* children, the finding that most of the child beggars are not from *Roma* origin is an interesting finding.

Age is also a determinant in the type of work. Data on prostitution is not mentioned in the IPEC study conducted in Turkey, however, studies by Küntay and Erginsoy (2000, 2005) provide information on teenage sex workers. Factors related to the patriarchal family structure are identified among the main causes for the girls to go to street and eventually engage in prostitution (Küntay and Erginsoy 2005, 16). Other factors are economic difficulties, migration and family disorganization (Küntay and Erginsoy 2005, 101). Unlike Romania, percentage of girls engaged in prostitution among working street children in general is not known. However, authors of the study, state that number of underage girls engaged in prostitution in Turkey is lower than other countries (Küntay and Erginsoy 2005, 16). There are also women/girls from Romania working as sex workers in especially Aksaray and Laleli district of İstanbul (Küntay and Erginsoy 2005, 129); however, whether these Romanian girls are under 18 years old or not. Family disorganization is a common factor and characteristics in both countries in the case of children engaged in prostitution.

In both countries, the children working on the street are exposed to certain risks. Noxious pollutants, car accidents, street gangs, sexual and physical violence (Küntay and Erginsoy 2005, 137), drug abuse, health and socialization risks are mentioned for Turkey (Kahveci et al. 1996, 57). The children in Romania are exposed to similar risks, however, references to adult pedophiles, state that 35% of the girls on the street are engaged in prostitution, and international child trafficking point to another face of street work (Alexandrescu 2002, 7). International trafficking issue is missing in Turkish studies; however, international studies indicate Turkey is involved either as a transition point or as a destination (Kane 2005, 1-12). In both countries, street work cause exhaustion because of working for long hours in hard jobs, physical injuries and psychological problems and high school drop out. Families tend not to be aware of the risks or they ignore due to difficulties in their living conditions. Usually the families tend to have a positive approach to child labour in general as they contribute to family budget and they hope that whatever it is, the work experience will help them in the future. Most of the children interviewed in both countries work voluntarily (or due to emotional blackmail by their parents) in order to contribute to their families. In both countries, some children mention that their parents (mostly fathers) force them to work (Akşit et al. 2001; Karatay et al. 2000a, 445; Alexandrescu 2002). Also the children in Romania tend to work on the street as an alternative for living together with their families which have violence and/or alcohol problems (Alexandrescu 2002, 6).

Generally speaking, majority of the children in Turkey somehow keep their connections with their families and houses. As extreme poverty is higher, maybe cultural dynamics play a role, effects of institutionalization are seen now as run aways to the streets, child abandonment and family organization is higher, street families (street children becoming pregnant on the street) and homeless families (families previously having a house, but lost their houses due to withdrawal of state housing subsidies) are much more, number of children living and working on the street in Romania might be claimed to be higher as a percentage among total number of working street children.

111

5.3.3 General Analysis and Comparison of Working Street Children in Turkey and Romania

The outcomes of the comparison of Turkey and Romania in terms of working street children in the context of New Poverty indicate that the global dynamics and integration of these countries to the international political economy at similar time periods lead to similarities while the national contexts account for the differences.

Specifically speaking, the communist ruling history of Romania has positive and negative implications on the well-being of children in the contemporary period. Free education, free health, subsidized housing, and maternal allowances directly affect well-being of the children, keeping under five mortality rates low, increasing literacy and life expectancy at birth. Thus, pre-1990 investments, particularly in health and education, increase children's living standards. However, institutionalization, high fertility rates due to pronatal policies and abandonment in relation to institutionalization and pronatal policies are also influential in the contemporary period. Collapse of the existing system and transition to the market economy and a Western type parliamentarian democracy meant a sudden deterioration in living standards increasing child poverty instantly.

Turkey, on the other hand, did not pass through such a period. Comparably fewer social expenditures and increased regional discrepancies due to the relatively fewer role of the state – combined with other factors such as economic crisis from time to time and far larger population which is more difficult to be served- caused Turkey to lag behind Romania in human development indices. However, on the other hand, continuity of the political system in Turkey helped make rapid progress especially in the last two decades. In terms of working street children, this contextual difference resulted in that while most of the working street children belong to the children working on the street category – those who work on the street but eventually go back to their houses and keep their connections with their families- in both countries; collapse of the system, institutionalization, abandonment, sudden increase in extreme poverty and sudden withdrawal of the existing free/subsidized social services resulted in higher incidence of street children in Romania. Existence of homeless families (families who lost their houses as housing supports are cut back, inflation and unemployment increased accompanying the decline in real wages) and street families (families composed out of street children who get pregnant on the street) are also relevant to the differences in national contexts. Besides, seemingly there are more children engaged in illicit activities in Romania because of higher extreme poverty.

Differences in the national contexts lead to differentiation in terms of child categories who under highest risks on the street. It might be claimed that girls are under higher risks in Romania, while boys comprise the risk group in Turkey as most of the girls tend to spend the night at home. Both the cultural structure and rate of extreme poverty are behind this gender difference. According to Table 5.1, in Turkey, children of the newly migrated crowded families in which mothers cannot speak Turkish are exposed to higher risks; while disorganized or single parent families, street families and homeless families' children are in the risk group in Romania. Garbage collection, based on the available data, involve highest risks in Turkey, while there is drug dealing, prostitution and trafficking in the reports of Romania which create significant risks for the children. Ethnicity is a determining factor in Romania in terms of job type, working street children categorization and risk factors. In the studies on Turkey, the ethnic dimension is referred in that majority of the mothers of working street children can not speak Turkish, rather they speak Kurdish and/or Arabic (Akşit et al. 2001, 63). In the study on İstanbul, language problem of mothers is reported to be lower (almost half of the mothers) (Karatay et al. 2000a, 442), however, there is still a connection between language skills of the mother and child's working on the street. In a study, two shoe-shine boys also state that "The children of our neighbours don't play with us. They say that we are dirty because we are Kurds...We speak Kurdish at home, but outside we speak only Kurdish" (Kahveci et al. 1996, 65). The problem with language skills might be interpreted as indicating low level of education which contributes to the poverty in the family as available jobs in majority require higher education. Language problem might also signify an ethnic dimension in terms of working street children in Turkey. Involvement of an ethnic dimension in both countries might be interpreted as an indicator of social exclusion in the emergence of working street children.

Table 5.2 provides a comparison between working street children in Turkey and Romania in terms of causes, scope and dimension. In terms of characteristics, most common age groups are closer; however, distribution of children in terms of sex is sharper in the Turkish case. School drop out rates are higher in Romania due to the sudden breakdown of the system and following economic crises.

Table 5.1 Risk Group

(Working Street Children who are under Highest Risks on the Street)

Characteristics	Turkey	Romania
Age	5 and over	5 and over
Gender	Boys	Girls[9]
Family	Crowded, newly migrated, lowly educated, mother cannot speak Turkish	Disorganized families, single parent families, street families, homeless families
Occupation	Garbage collection[10]	Drug dealing, prostitution, trafficking
Ethnicity	Kurdish/Arabic speaking	*Roma*

In terms of health, HIV/AIDS incidence is found among Romanian working street children, although again this might be explained with lack of data for Turkey, it can also be argued that higher number of street children and street families might be responsible for spread of the disease. Origin of working street children in Turkey is Southeastern and Eastern Anatolia due to partly social unrest, and partly economic difficulties and regional development discrepancies. In Romania, the Northeastern region is the origin of most children because of the deteriorated agricultural structure on which the regional economy is established. The national context is also reflected upon the dynamics behind emergence and expansion of working street children. While migration and earthquake emerge as country-specific factors in Turkey, in Romania family

---

[9] Prostitution and child trafficking seems to be more common in the Romanian case. In Turkey, among children working on the street and having contact with their families, the girls are usually accompanied by older brothers and they are usually pulled back after a certain age. The fact that living on the street is less common in Turkey contributes to this issue. On the other hand, among children living on the street, there is incidence of prostitution (Küntay and Erginsoy 2005). Therefore, Table 5.1 is referring to the general population of children engaged in street work, majority of whom continue to have contact with families.

[10] Prostitution is not mentioned in Table 5.1, since majority of working street children are not living permanently on the street, and therefore, garbage collection emerges as the most common and risky task on the street.

disorganization (violence, alcohol and drug addiction, disintegration of family due to death or divorce) comes forward.

Table 5.2 Working Street Children in Turkey and Romania

Characteristics	Turkey	Romania
Age[11]	9-11	12-14
Gender	Boys over 85%, Girls around 10-15%	Boys around 70%, Girls around 30%
Education	School drop out over 30% in general, boys usually go to school, girls are in majority among those who never attended school, girls are the first to be taken back from school	School drop out over 60% in general, more boys go to school, girls are in majority among those who never attended school, girls are the first to be taken back from school. School drop out rates and attainment vary in accordance with being a *Roma* child or not.
Living conditions	Mostly live with family, work on the street; some both live and work	Majority works on the street but keep connections with families, yet higher incidence of living on the street[12]
Occupation	Shoe polishing, selling petty commodities, scavenging, car cleaning, loading and unloading, garbage collection and separation	Begging, car washing, garbage collection, loading and unloading, household work, selling petty commodities, stealing and prostitution.
Ethnic dimension	Kurdish-Arabic speaking in majority	*Roma* children
Sibling number	Crowded	Crowded, almost always over 3
Mother's education	Mostly illiterate, cannot speak Turkish	Depend on ethnicity. Majority speaks Romanian language, but illiterate if *Roma*. If not, literate but lowly educated
Father's education	Mostly literate, but low	Literate, low educated.

---

[11] The most common age group.
[12] There is the case of previously institutionalized children, child abandonment, street families, homeless families.

	educated	
Parental Employment	Mother housewife, father unemployed or informal sector employment	Mother housewife, father unemployed or informal sector employment
Health	Exhaustion, injuries, depression, drug addiction (glue sniffing)	Exhaustion, injuries, depression, HIV/AIDS incidence, drug addiction
Most Common Places	İstanbul and Marmara Region, big regional cities (Adana, Diyarbakır, İzmir etc.)	Bucharest and Ilfov region and big regional cities (Iaşi and Craiova)
Origin	Eastern and Southeastern Anatolia	Northeastern Romania
Causes	Poverty, migration, earthquake, low education level of families, legislative deficiencies	Poverty, family disorganization, low education level of the families, legislative deficiencies

In the two countries, poverty stands for the shared dynamic behind working street children. Low education level of the families might be evaluated in relation to poverty. Thus, an analysis of the similarities might take poverty as the starting point. As explained in the introduction chapter and Chapter I, increasing pace of the globalization process within the framework of neo-liberal paradigm resulted in multiple transformations in politics, economy and society. The states shrank and their roles in economy diminished. Market structure changed from an industrial production system to an overwhelmingly service sector economy in which subcontracting, informal market, deregulation and labour disorganization have become dominant. Majority of the social provisions and expenditures are cut in this economic political system. This multifaceted process resulted in transformation of poverty and emergence of New Poverty. In this new period, relative poverty increased despite falling absolute poverty levels, and inequality and social exclusion became important in creating winners and losers. Thus, increased relative poverty combined with withdrawal of social policies leading to loss of state support, change in the family leading to loss of familial support and structural change in the market leading to loss of available jobs resulted in increased tendency to work on the street. This process helps explain why street children emerged and why they emerged in an era of high economic growth. Moreover, integration of Turkey to international political economy through structural adjustment programmes; and Romania through a systemic change and again the structural adjustment programmes which followed the transition to a market economy both in early 1990s is useful in explaining why the problem emerged in two so different countries in close time periods.

CHAPTER VI

CONCLUSION

Working street children have emerged at a period when high economic growth rates are claimed to be reached; many commodities, which were previously considered as luxury items, have become common and technological innovations have been making rapid progress. What has driven me to study on working street children was such a starting point. Neither poverty nor child labour is specific to the contemporary decades. However, working on the street, for children, as a permanent economic activity and rapid expansion of the number of children engaged in tasks on the street might be considered as particular to the early 1990s.

What has changed so that the children and their families are directed towards street work, not to a more traditional form of child labour? How can we explain emergence of working street children particularly in 1990s, although the same dates also refer to one of the highest national and global economic performance periods ever? Why do we come across with the same incidence of child labour, namely working street children, in quite distinct countries in similar time periods?

In order to provide answers to these research questions, a two-fold analysis is conducted, one on poverty, the other being on the international dimension. In contemporary societies, psychological value of the children prevails over their economic value under normal conditions. Thus, poverty is the principal motive behind child work. What has changed with poverty, and subsequently with child poverty, is examined in order to understand what has changed with child labour, as child labour is historically among the family survival strategies against poverty. The second analysis focuses on the international dimension in order to evaluate national contexts more comprehensively.

Throughout the study, comparative historical methodology is employed. Turkey and Romania is compared from 1990s onwards, paying specific attention to the political and economic change in these years. Both the country-specific and international factors are considered for understanding why working street children emerged in both countries despite differences in the national context. The transformation of poverty in its historical course and the dynamics behind differences and similarities are provided through this methodology.

The comparison of Turkey and Romania in terms of causes and characteristics of working street children from the perspective of our research questions indicate two points. Firstly, the differences between two countries mostly depend on two countries' having different economic, political and social contexts and similarities, particularly in the political economic process that led to emergence of working street children, point to Turkey's and Romania's increased integration to the international environment. Secondly, the comparison of two countries reveals that emergence of working street children can not be solely explained on the basis of poverty in the absolute sense. Therefore, these two general conclusions reached through the historical comparative analysis of Turkey and Romania lead us to New Poverty as it is discussed in the theoretical chapter.

In the theoretical chapter, New Poverty is defined as a new form of poverty, in which absolute poverty referring to no income and low income play role, nevertheless relative poverty in the sense of being deprived from the available resources and opportunities in the society which was accessed by the rest of the society is more influential (Gordon 2000, 49-50). Why could some sections of the society "have" and why some others are deprived is explained by the existence of inequality and social exclusion (Blakemore 2003, 80; Esping-Andersen 1999, 44). Besides these characteristics, New Poverty is claimed to refer to a condition in which global forces are more persistent and poverty might have similar reflections in different societies through the integration of these societies to the international political economy (UN 1995, 41).

During the study, it is seen that poverty is the main cause for the emergence of working street children in Turkey and Romania. In both countries, in the families of working street children, the studies show that majority of male households are unemployed or employed in the informal sector with low salaries (Akşit et al. 2001, Alexandrescu 2002). Thus, it can still be argued that despite the decreases in absolute poverty in both countries, there is still problem with income in the sense of market transfers. However, the findings that majority of the children are coming from Kurdish (and/or Arabic) speaking families in Turkey (Akşit et al. 2001; Kahveci et al. 1996, Altıntaş 2003) and from *Roma* families in Romania (Alexandrescu 2002; Zamfir 2002; Save the Children Romania and IPEC 2003) indicate existence of social exclusion. Economic growth might be achieved in these societies, yet it is not distributed equally in these societies, particularly socially excluded groups, especially the recently migrated groups could not benefit from the economic performance.

Existence of relative poverty, social exclusion and inequality dimensions of New Poverty in the Turkish and Romanian contexts leads us also to the neo-liberalism and globalization interaction, discussed in the theoretical chapter. In the theoretical chapter, it was mentioned that national economies are increasingly opened to the global forces through the technological advances and through the relations with the international monetary organizations. Neo-liberal paradigm has provided a political economic project to the governments through structural adjustment programmes in which public expenditures would be decreased in order to be more competitive and also to pay their debts (Nissanke and Thorbecke 2005). Turkey and Romania are both considered as medium human development countries (UNDP 2005). Therefore, families living in extreme poverty and children who has to work for long hours, exposed to several physical and mental dangers on the street, some of them spending the night on the street (together with their families in the Romanian case, Zamfir 2002) invokes the question about policy choices. Previously, it was also mentioned that Turkey's human development performance in terms of children is not compatible with its economic performance (Buğra and Keyder2005). Thus, outcomes of the comparison of two countries in terms of child poverty signifies a need to question social policy considerations. Neo-liberal paradigm as the dominant understanding behind reform measures taken by both governments have resulted in withdrawal or cutback of social services particularly in the field of education and health in the social sphere. In the economic sphere, as the economic pattern has shifted to service sector with low paid, temporary and part time jobs for low skilled and low educated workers has caused increase in poverty of the household. The working street children of Turkey and Romania are directed to the streets in such a context of decreased state and family support. Therefore, in my opinion, social policy cutbacks play more roles in the emergence and expansion of working street children in both countries than the role of income poverty. For this reason, while emphasizing New Poverty as the principal cause of working street children, not only absolute poverty, but also increase of relative poverty on the basis of the social policy cutbacks is meant. I believe if, at least, public services on education and health which affect children more directly could be preserved, the scope of working street children would not be as extended as now in both countries. The fact that extension of compulsory primary education in both countries helped withdrawal of a number of

121

children from the streets back to schools (Gündüz Hoşgör 2004; Save the Children Romania 2001) might be considered to strengthen this argument.

Involvement of decreased state support combined with an international dimension through repercussions of global dynamics and neo-liberal paradigm in the emergence of working street children in both countries imply that policies on the macro level are important for children. Projects and programmes targeting specific groups in a particular area, city or neighbourhood might be quite successful in returning children from the street to the school. Cash transfers and income generation for the families of working street children at a particular district might also be successful. Particularly, activities of NGOs are significant from this perspective. Returning each child to the school is valuable. However, eradicating street work and other forms of child labour might be achieved through policies on the national level. In both countries, there are many other children who live in poverty and who are under the risk of beginning to work and/or live on the street. Unless the children on the risk of beginning to work and/or live on the street are protected through national social policies, efforts might remain limited to individuals or groups. Developing policies specifically targeting children, providing support in terms of health, education and housing on the national level, might be a preventive measure.

For the children already working/living on the street, again devising national policies for the children is important. Success of the extension of compulsory primary education in decreasing number of the children on the street in both countries indicate importance of education in the policies for these children. Legal regulations directly addressing children on the street and persons abusing these children and enforcement of them might be suggested particularly as a measure in the cases where there is organized exploitation of working street children. Facilities regarding health, education and accommodation of the children provided by the national institutions and NGOs in both countries provide significant services to the working street children. Increasing the number of similar centers and institutions and improving the conditions of the existing ones within the context of a broader national policy for working street children might improve the services to the working street children.

On the whole, in both countries, working street children is a considerably new problem. The reports indicate that illicit activities compose a small percentage of the tasks the children are engaged with. It is even reported that children on the street are not deviant initially. Ongoing efforts have been successful for part of the children in both countries, therefore, mainstreaming the policies into the national policies might change the lives of these children permanently and save lives of many others who are at the risk of becoming working street children.

Continuing the studies and research, in this context, is also important as these studies might provide the necessary basis for the future policies. Further and detailed research on the gender dimension, illicit activities such as drug dealing and prostitution, exploring incidence of organized crime involving the children on the street and specific attention on the ethnic dimension might provide more detail understanding on the issue of working street children. Since migration and social policies stand as important causes behind emergence and expansion of working street children in both countries, further research and consequent policy proposals are of particular importance.

REFERENCES

Adumitracesei, I.D. and Niculescu, N.G. (1999). *Post-Socialist Romanian Economy*. Bucharest: Editura Economica.

Akşit, B., Karancı, N. and Gündüz Hoşgör, A. (2001). *Turkey: Working Street Children in Three Metropolitan Cities: A Rapid Assessment*. Geneva: IPEC.

Alexandrescu, G. (2002). *Romania, Working Street Children in Bucharest: A Rapid Assessment*. Geneva: IPEC.

Altıntaş, B. (2003). *Mendile, Simite, Boyaya, Çöpe...Ankara Sokaklarında Çalışan Çocuklar*. İstanbul: İletişim.

Atauz, S. (1989a). "Child Labour and Street Children in Turkey". *Status Report presented to UNICEF*. Ankara.

Atauz, S. (1989b). "Sanayi Kesiminde Çalışan Çocuklar" in Konanç, E., Gürkaynak, İ and Egemen, A. (eds.). Çocuk İstismarı ve İhmali. Çocukların Kötü Muaemeleden Korunması 1. Ulusal Kongresi (12-14 Haziran). Ankara.

Atauz, S. (1990a). *Ankara ve Şanlıurfa'da Sokak Çocukları*. Ankara: UNICEF.

Atauz, S. (1990b). *Street Children in Ankara*. Ankara: UNICEF.

Atauz, S. (1995). *İzmir Sokaklarında Yaşayan/Çalışan Çocuklar*. Sokak Çocukları Koruma Derneği: İzmir.

Atauz, S. (1996). "Sokak Çocuklarının Kenti" in Komut, E. (ed.) *Diğerlerinin Konut Sorunları: Habitat II*. Ankara: Mimarlar Odası Yayınları.

Atauz, S. (1997). *Diyarbakır Sokak Çocukları Araştırması*. Ankara: Uluslararası Lions, MD 118 U Yönetim Çevresi.

Balamir, A. (1982). "Türkiye'de İmalat Sanayinde Çocuk İşgücü". *Nüfusbilim Dergisi*, Hacettepe University, Vol.4.

Basu, K. And Tzannatos, Z (2003). "The Global Child Labor Problem: What do we Know and what can we Do?". *The World Bank Economic Review*. Vol. 17, No.2.

Becker, S. (1997). *Responding to Poverty: The Politics of Cash and Care*. London: Longman.

Blakemore, K. (2003). *Social Policy: An Introduction*. Buckingham: Open University Press.

Boratav, K., Yeldan, A.E. and Köse, A.H. (2000). "Globalization, Distribution and Social Policy: Turkey, 1980-1998". *Center For Policy Analysis Working Paper Series No. 20*. New York.

Bradburry, B., Jenkins, S.P. and Micklewright, J. (eds.) (2001). *The Dynamics of Child Poverty in Industrialized Countries*. Cambridge: Cambridge University Press.

Bradbury, B. and Jännti, M. (2001) "Child Poverty across the Industrialized World: Evidence from the Luxembourg Income Study" in Vleminckx, K. and Smeeding, T.M. (eds.) (2001). *Child Well-being, Child Poverty and Child Policy in Modern Nations: What Do We Know?* Bristol: The Policy Press.

Bradshaw, J. (2000). "Child Poverty in Comparative Perspective" in Gordon, D. and Townsend, P. (eds.) (2000). *Breadline Europe: The Measurement of Poverty*. Bristol: The Policy Press.

Bradshaw, J. and Barnes, H. (1999). "How do Nations Monitor the Well-being of their Children?". *Luxembourg Income Study Child Poverty Conference*, September 30-October 2, 1999, Luxembourg.

Buğra, A. and Keyder, Ç. (2003). *New Poverty and the Changing Welfare Regime of Turkey*. Ankara: UNDP.

Buğra, A. and Keyder, Ç. (2005). *Poverty and Social Policy in Contemporary Turkey*. İstanbul: Boğaziçi University Social Policy Forum.

Burchardt, T. (2000). "Social Exclusion: Concepts and Evidence", in Gordon, D. and Townsend, P. (eds.) (2000). *Breadline Europe: The Measurement of Poverty,* Bristol: The Policy Press.

CASPIS (Anti-Poverty and Social Inclusion Commission in Romania) (2002). *National Anti-Poverty and Social Inclusion Plan, Poverty Dynamics 1995-2001.* Bucharest: CASPIS.

Christopher, K., England, P., McLanahan, S., Ross, K. and Smeeding, T.M. (2001). "Gender Inequality in Poverty in Affluent Nations: The Role of Single Motherhood and the State" in Vleminckx, K. and Smeeding, T.M. (eds.) (2001). *Child Well-being, Child Poverty and Child Policy in Modern Nations: What Do We Know?* Bristol: The Policy Press.

Commission of the European Communities (2002). *Communication from the Commission to the Council and the European Parliament: Roadmaps for Bulgaria and Romania.* Brussels: EU.

Cornia, G.A. (2000). "Inequality and Poverty in the Era of Liberalisation and Globalisation". *UNU/WIDER Discussion Paper,* in Cornia, G. A. (ed.) (2004). *Inequality, Growth, and Poverty in an Era of Liberalization and Globalization.* Oxford: Oxford University Press.

Cornia, G.A. (2001). *Impact of Liberalisation and Globalization on Income Inequality in the Developing and Transitional Economies.* Florence: University of Florence.

Cumhuriyet (10.02.2005), "Sokaklarda 'Büyüyen' Tehlike".

Çocuk Vakfı (2000). *Hakları Çalınmış Çocuklar Raporu.* İstanbul: Çocuk Vakfı.

Daniel, P. and Ivatts, J. (1998). *Children and Social Policy.* London; New York: Palgrave.

Dansuk, E. (1997). "Türkiye'de Yoksulluğun Ölçülmesi ve Sosyo-ekonomik Yapılarla İlişkisi". *Expertise Thesis.* Ankara: SPO.

Dayıoğlu, M. and Gündüz Hoşgör, A. (2004). *IPEC Experience in Turkey.* Ankara: IPEC.

Dutu, M. (2002). *Study on the Current Romanian Legislation Regarding Child Labour and Recommendations on the Necessary Amendments for its Approximation with the International Regulations.* Bucharest: IPEC and Government of Romania.

127

Dziewiecka-Bokun, L. (2000). "Poverty and the Poor in Central and Eastern Europe" in Gordon, D. and Townsend, P. (eds.) (2000). *Breadline Europe: The Measurement of Poverty*, Bristol: The Policy Press.

Ecevit, Y. (1998). "Küreselleşme, Yapısal Uyum ve Kadın Emeğinin Kullanımında Değişmeler" in Ferhunde Özbay (ed.) (1998). *Küresel Pazar Açısından Kadın Emeği ve İstihdamındaki Değişmeler: Türkiye Örneği*. İstanbul: İnsan Kaynağını Geliştirme Vakfı.

Emmerson, C., Frayne, C. and Goodman, A. (2002). "How much would it Cost to Increase UK Health Spending to the European Union Average?". *Research Report*. Institute for Fiscal Studies: London.

Erder, S. (2001). "Yeni Yoksul(n)luklar Üzerine". *Sosyal Demokrat Değişim*. No.19.

Erder, S. (2004). "'Yeni' Yoksulluk ve 'Yeni' Modeller". *Kızılcık*. Mart/Nisan 2004.

Ertürk, Y. (1994). *Patterns of Child Labour in Rural Turkey*. Ankara: ILO.

Erman, T. (2003). "Poverty in Turkey; The Social Dimension" in World Bank (2003). *Turkey: Poverty and Coping after Crises Volume 2*. Report No.24185-TR. Washington DC: World Bank.

Ertürk, Y. and Dayıoğlu, M. (2004). *Gender, Education and Child Labour in Turkey*. Geneva: IPEC.

Esping-Andersen, G. (1999). *Social Foundations of Postindustrial Economies*. Oxford; New York: Oxford University Press.

Esping-Andersen, G. (ed.) (1996). *Welfare States in Transition: National Adaptations in Global Economies*. London: Sage Publications, published in association with the United Nations Research Institute.

EU (2004a). *2003 Annual Report on Turkey's Progress towards Accession*. Brussels: EU.

EU (2004b). *2003 Regular Report on Romania's Progress towards Accession*. Brussels: EU.

European Commission Directorate General Enlargement Information and Interinstitutional Relation (2003). *Enlargement of the European Union, Guide to the Negotiations Chapter by Chapter.* Brussels: EU.

European Commission (2004). "Eradicating the Worst Forms of Child Labour in Turkey". *Project Fiche.* Project No. 0403.04.

European Parliament Resolution on Romania's Application for Membership of the European Union and the State of Negotiations (COM(2000) 710 - C5-0610/2000 - 1997/2172(COS)), Article 49.

European Union and Romanian Government (2003). "Romania's Children: Their Story". *Information Booklet.* Bucharest: Government of Romania.

Eurostad (2004). *Population and Social Conditions, European Communities.* Catalogue n. KS-NK-04-02-EN-N.

Fafo Institute for Applied International Studies (2004). *Manual for Rapid Assessment: Trafficking in Children for Labour and Sexual Exploitation in the Balkans and Ukraine.* Geneva: IPEC.

Fazlıoğlu, A. (2002). "'An Urban Reality' Children Working in the Street". *The Hope.* Ankara, No.2.

Fişek Institute. http://www.fisek.org.tr (access date: 22.05.2006).

George, V. and Wilding, P. (2002). *Globalization and Human Welfare.* New York: Palgrave.

Ghinararu, C. (2004). "Child Labour in Romania". *Discussion Paper.* Bucharest: Ro Media.

Goldson, B., Lavalette, M. and McKechnie, J. (eds.) (2002). *Children, Welfare and the State.* London; Thousand Oaks; New Delhi: Sage.

Gordon, D. (2000). "Measuring Absolute and Overall Poverty" in Gordon, D. and Townsend, P. (eds.) (2000). *Breadline Europe: The Measurement of Poverty.* Bristol: The Policy Press.

Gordon, D. and Townsend, P. (eds.) (2000). *Breadline Europe: The Measurement of Poverty.* Bristol: The Policy Press.

Government of Romania and UN Romania (2003). *Millenium Development Goals.* Bucharest: RoMedia.

Government of Romania (2005). *Government Programme 2005-2008.* Bucharest: RoMedia.

Government of Romania, National Authority for the Protection of the Child and Adoption (2001). *Government Strategy Concerning the Protection of the Child in Difficulty (2001-2004).* Bucharest: RoMedia.

Government of Romania National Authority for Child Protection and Adoption (2003). *Child Protection between Results and Priorities for the Future.* Bucharest: RoMedia.

Gregorian, R., Hura-Tudor, E. and Feeny, T. (2003). "Street Children and Juvenile Justice in Romania". *International Project of Asociatia Sprijinirea Integrarii Sociale (ASIS) and The Consortium for Street Children.* Bucharest: ASIS.

Gündüz Hoşgör, A. (2001). "Convergence between Theoretical Perspectives in Women-Gender and Development Literature regarding Women's Economic Status in the Middle East". *METU Studies in Development.* Vol.28, Nos. 1-2.

Gündüz Hoşgör, A. (2004). *"Good Practice" Applications: Guidelines for Action against Children Working on the Streets as a Worst Form of Child Labour in Turkey.* Ankara: IPEC.

Gündüz Hoşgör, A., Karabıyık, E., Çetinkaya, Ö., and Sargın, H.C. (2005). *Sokaktan Umuda: Başarı Öyküleri. Sokakta Çalışan Çocuklar Sorununun Çözümüne Yönelik Çalışmalar ve Yöntemsel Rehber.* Ankara: ILO.

Holman, R. (1978). *Poverty: Explanations of Social Deprivation,* London: Martin Robertson.

ILO (1973). *Minimum Age Convention No. 138.*

130

Konanç, E. (1991). *Ankara'da Sokakta Çalışan Çocuklar Konusunda Araştırma Bulguları*. Ankara: ILO.

Konanç, E. (1996). "Çalışan Çocuklara İlişkin Uygulamalar ve Hukuki Düzenlemeler" in Kahramanoğlu, E. *Türkiye'de Çalışan Çocuklar Sorunu ve Çözüm Yolları*. Ankara: H.Ü. Sosyal Hizmetler Yüksek Okulu and Friedrich-Naumann Vakfı.

Kurt, S. (09.02.2001). "Terk Edilen Çocuk Sayısı Artıyor". *Zaman*.

Küntay, E. and Erginsoy, G. (2000). "Ticari Seks İşçisi olarak Sömürülen Çocukların Görüşleri ve Gereksinimleri" in *I. Istanbul Çocuk Kurultayı-Bildiriler Kitabı*. İstanbul: İstanbul Çocukları Vakfı.

Küntay, E. and Erginsoy, G. (2005). *İstanbul'da On Sekiz Yaşından Küçük Ticari "Seks İşçisi" Kız Çocuklar*. İstanbul: Bağlam.

Küntay, E., Erginsoy, G. and Yılmaz, E. (1998). *Selpağa Gitmek: Sokakta Çalışan Çocuklar. Pilot Araştırma Raporu*. İstanbul: Provincial Governate of Istanbul.

Luxembourg Income Study (1994), at http://lis.ceps.lu. (access date:10.04.2006).

Mahoney, J. (2004). "Comparative-Historical Methodology". *Annual Review of Sociology*. No.30. Online publication. http://arjournals.annualreviews.org/doi/abs/10.1146/annurev.soc.30.012703.110507;jsessionid=iP LNti6kiKhcj6EppU?cookieSet=1&journalCode=soc. (access date: 01.06.2006).

Micklewright, J. (2002). "Social Exclusion and Children: A European View for a US Debate". *CASEpaper No.51*. London.

Ministry of Labour and Social Security Department General of Labour Division of Child Labour (2005). *Türkiye'de Çocuk İşçiliği, Sorun Bizim: Bilgilendirme Materyali, Kitap 1*. Ankara: MLSS.

Mother and Child Care Institute "Alfred Rusescu", UNICEF and Ministry of Labour Social Solidarity and Family, National Authority for Child Rights Protection (2005). *The Situation of Child Abondonment in Romania*. Bucharest: UNICEF.

Nastase, A. (ed.) (2001). *Romania and the Future of Europe.* Bucharest: Monitorul Oficial.

National Institute of Statistics (2003). *The 2003 Turkey Demographic and Health Survey.* Ankara: NIS.

Neuman, W.L. (2000). *Social Research Methods: Qualitative and Quantitative Approaches.* Boston: Allyn and Bacon.

Niculescu, N.G. and Adumitracesei, I.D. (2000). *Post-Socialist Romania Confronted with Underdevelopment.* Bucharest: Editura Economica.

Niculescu, N.G. and Adumitracesei, I.D. (2001). *Romania on the Way to European Economic Integration.* Bucharest: Editura Economica.

Nissanke, M. and Thorbecke, E. (2005). "The Impact of Globalization on the World's Poor: Transmission Mechanisms". *WIDER Jubilee Conference in Helsinki,* June 17-8, 2005.

Oto, R., Melikşah, E. and Geter, R. (n.d.). *Diyarbakır Sokaklarında Çalışan Çocukların Özelliklerine İlişkin Bir Çalışma.* Research report.

Özbay, F. (1991). "Türkiye'de Kadın ve Çocuk Emeği". *Toplum ve Bilim.* No: 53. İstanbul.

Özbay, F. (1998). "Türkiye'de Evlatlık Kurumu: Köle mi Evlat mı?" in *International Conference on History of Turkish Republic: A Reassesment, Volume II. Economy, Society and Environment.* Ankara: Tarih Vakfı Yayınları.

Özcan, Y.Z. (2003). "Measuring Poverty and Inequality in Turkey" in *Turkey: Poverty and Coping After Crises Volume II: Background Papers,* Washington: World Bank.

Penton, R. (1999). "Romania Country Report". *Children and Residential Care Alternative Strategies,* May 3-6 1999, Stockholm.

Phipps, S. (2001). "Values, Policies and Well-being of Young Children in Canada, Norway and the United States" in Vleminckx, K. and Smeeding, T.M. (eds.) (2001). *Child Well-being, Child Poverty and Child Policy in Modern Nations: What Do We Know?* Bristol: The Policy Press.

Redmond, G., Schnepf, S.V. and Suhrcke, M. (2002). "Attitudes to Inequality after Ten Years of Transition". *Innocenti Working Papers No. 88.* Florence: Innocenti Research Center.

Romanian Government (2003). "In the Interest of the Child". *Newsletter,* No.2, Bucharest: Government of Romania.

Romanian National Institute of Statistics and ILO (2003). "Survey on Children's Activity in Romania". *Country Report.* Bucharest: RNIS.

Save the Children Romania (2002). *Report by Salvaţi Copiii România (Save The Children Romania) to the UN Committee on the Rights of the Child – Geneva concerning the Second Periodical Report by the Romanian Government on the Interval 1993 – 2002.* Geneva: Save the Children.

Save the Children Romania (2002). *Rroma Working Children and their Families.* Bucharest: Ro Media.

Save the Children Romania (2003). "Child Trafficking in Central, Southeastern Europe and Baltic Countries". *Regional Report.* Bucharest: Save the Children.
Save the Children Romania and IPEC (2003). *Baseline Surveys on Working Street Children in Bucharest, Iaşi and Craiova.* Bucharest: IPEC.

Save the Children Romania (2003). *Study on Child Labour in Romania.* Bucharest: IPEC.

Save the Children Romania (2005). *Working Street Children.* Bucharest: Save the Children.

SHÇEK, DİE and UNICEF Turkey (2005). *Katılımlı Eylem Araştırması: Sokakta Yaşayan ve Çalışan Çocuklar.* Ankara: UNICEF.

SHÇEK (2005a). *Çocuklara Verilen Hizmetler.*
http://www.shcek.gov.tr/portal/dosyalar/hizmetler/cocuk/hizm_cogem.asp   (access   date:
01.07.2006).

SHÇEK (2005b). *Ocak 2005 İtibariyle Çocuk ve Gençlik Merkezlerinden Hizmet Alan Çocuk Sayıları.*

http://www.shcek.gov.tr/portal/dosyalar/hizmetler/cocuk/sokak_cocuk/2004_cogem.asp (access date: 01.07.2006).

SHÇEK (2006). *Çocuk ve Gençlik Merkezleri.*
http://www.shcek.gov.tr/portal/dosyalar/turkiye/tablo_cogemx.asp (access date: 03.07.2006).

Southeastern Anatolian Project (GAP) (2000). "Diyarbakır Kentinde Sokakta Çalışan Çocukların Rehabilitasyonu Projesi". *Information Sheet No.28.* Ankara.

State Institute of Statistics (1999). *Child Labour in Turkey 1999.* Ankara: SIS.

State Institute of Statistics (2004). "2002 Yoksulluk Çalışması Sonuçları". *DİE Haber Bülteni.* Ankara: SIS.

State Planning Organization (2002). *İller ve Bölgeler İtibariyle Çeşitli Göstergeler.* Ankara: SPO.

Titmuss, R. M. (edited by Brian Abel-Smith and Kay Titmuss) (1974). *Social Policy: An Introduction.* London: Allen and Unwin.

Topal, Ç. (June 2004). *An Inquiry into Rural-Development Nongovernmental Organizations in Turkey: Degree of Institutionalization and Socio-Economic Characteristics of the Employees.* Unpublished Thesis. Ankara: METU.

Udry, C. (2003). "Child Labour". *Yale University Economic Growth Center Discussion Paper No. 856.* http://www.econ.yale.edu/growth_pdf/cdp856.pdf (access date: 28.05.2006).

UN (1989). *Conventions on the Rights of the Child.*

UN (1989). *Conventions on the Rights of the Child.*

UN (1995). *Report of the World Summit for Social Development.* Copenhagen, 6-12 March 1995, A/CONF.166/9, 19 April 1995.

UN (2003). *The Millenium Development Goals.* Bucharest: UN.

UNDP (1999). *Human Development Report 1999: Human Development in the Age of Globalization.* New York: Oxford University Press.

UNDP (2002). *A Decade Later: Understanding the Transition Process in Romania, National Human Development Report Romania 2001-2002.* Bucharest: UNDP Romania.

UNDP (2005). *Human Development Report 2005.* Geneva: UNDP.

UNDP (2005). *Human Development Report 2005.* Geneva: UNDP.

UNDP Romania (1998). *National Human Development Report 1997.* Bucharest: UNDP.

UNDP Romania (1998). *National Human Development Report 1997.* Bucharest: UNDP.

UNDP Romania (2005). *Millenium Development Goals in Romania.* Bucharest: UNDP.

UNICEF (1999). "After the Fall: The Human Impact of Ten Years in Transition", *The MONEE Project.* Florence: UNICEF Italy.

UNICEF (2002). *Çalışan Çocuklar Donör Haritalama Çalışması: I. Kuruluşlar, II. Program ve Projeler, III. Olası Donör Kuruluşlar Raporu.* Ankara: UNICEF.

UNICEF (2004). *The State of the World's Children 2005: Childhood under Threat.* New York: UNICEF.

UNICEF (2005). *The State of the World's Children 2006: Excluded and Invisible.* New York: UNICEF.

UNICEF (2006). *Çocuk Yoksulluğu'nun Önlenmesi.* Brochure. Ankara: UNICEF.

UNICEF Innocenti Research Center (1997). "Children at Risk in Central and Eastern Europe: Perils and Promises". *The MONEE Project No.4.* Florence: Innocenti Research Center.

UNICEF Innocenti Research Center (2000). "A League Table of Child Poverty in Rich Nations". *Innocenti Report Card Issue No.1*, June 2000. Florence: Innocenti Research Center.

UNICEF Innocenti Research Center (2004). *Innocenti Social Monitor 2004: Economic Growth and Child Poverty in the CEE/CIS and the Baltic States.* Florence: Innocenti Research Center.

UNICEF Romania (2005). *Information Brochure.* Bucharest: UNICEF.

UNICEF Turkey (2005). "The Race is on! Turkey, Children and MDGs". *Say Yes.* Autumn 2005.

Union of Health and Social Service Labourers (2006). *Public Release.* http://www.bianet.org/2005/11/25/70690.htm (access date: 31.03.2006).

Vilnoiu, M. (2000). "Poverty and Social Exclusion in Romania". *Presentation Paper.* Dale Carnegie Training.

Vleminckx, K. and Smeeding, T.M. (2001) "Ending Child Poverty in Industrialized Nations" in Vleminckx, K. and Smeeding, T.M. (eds.) (2001). *Child Well-being, Child Poverty and Child Policy in Modern Nations: What Do We Know?* Bristol: The Policy Press.

Vleminckx, K. and Smeeding, T.M. (eds.) (2001). *Child Well-being, Child Poverty and Child Policy in Modern Nations: What Do We Know?* Bristol: The Policy Press.

Westhof, D. (1997). *Flow Model Institutionalised Children In Romania And The Determining Variables.* Florence: UNICEF.

World Bank (2002). *Transition: The First Ten Years, Analysis and Lessons for Eastern Europe and the Former Soviet Union.* Washington: WB.

World Bank (2003). *Turkey: Social Risk Mitigation Project/Loan. Report No: 22510-TU.* Washington DC: World Bank.

World Bank (2005). *Romania Country Assistance Evaluation.* Report No. 32452. www.wds.worldbank.org/external/default/WDSContentServer/IW3P/IB/2005/06/20/000012009_20050620092117/Rendered/PDF/324520rev.pdf) (access date: 28.05.2006.

World Bank (2006a). *World Development Report 2006: Equity and Development.* Oxford: World Bank and Oxford University Press.

World Bank (2006b). *Turkey Labour Market Study Summary.* Report No. 33254-TR. Washington DC: Poverty Reduction and Economic Management Unit of World Bank.

World Bank (2006c). *Turkey Labour Market Study Full Report.* Report No. 33254-TR. Washington DC: Poverty Reduction and Economic Management Unit of World Bank.

Yener, S. and Kocaman, T. (1989). "Çocuk Nüfusun Demografik Özellikleri" in *Türkiye'de Çocuğun Durumu.* Ankara: State Planning Institute and UNICEF Representation in Turkey.

Zamfir, C. (1997). "Poverty in Romania". *Discussion Paper.* www.warwick.ac.uk/fac/soc/complabstuds/russia/Poverty_in_Romania.doc– (access date: 12.03.2006).

Zamfir, C. (2002). *Poverty in Romania.* Bucharest: UNDP.

Zamfir, C. and Zamfir, E. (1996). *Children at Risk in Romania: Problems Old and New.* Florence: UNICEF Innocenti Research Center.

Zeytinoğlu, S. (1989). *Working Children and Street Children.* İzmir: Aegean University.

APPENDIX

ILO AGREEMENTS CONCERNING CHILDREN

Among the ILO Agreements accepted at various dates since 1937, those concerning children are as follows:

ILO Agreement No. 15
Acceptance date by ILO 1921

ILO Agreement No. 45
Acceptance date by ILO 1935

ILO Agreement No. 58
Acceptance date by ILO 1936

ILO Agreement No. 59
Acceptance date by ILO 1937

ILO Agreement No. 77
Acceptance date by ILO 1946

ILO Agreement No. 123
Acceptance date by ILO 1965

ILO Agreement No. 138
Acceptance date by ILO 1965

ILO Agreement No. 182
Acceptance date by ILO 1998

Printed by
Schaltungsdienst Lange o.H.G., Berlin